Collins

Revision

NEW GCSE SCIENCE

Chemistry

or AQA A Higher

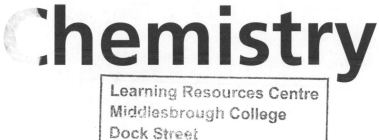
Author: Rob Wensley

Revision guide +
Exam practice workbook

William Collins' dream of knowledge for all began with the publication of his first book in 1819. A self-educated mill worker, he not only enriched millions of lives, but also founded a flourishing publishing house. Today, staying true to this spirit, Collins books are packed with inspiration, innovation and practical expertise. They place you at the centre of a world of possibility and give you exactly what you need to explore it.

Collins. Freedom to teach

Published by Collins
An imprint of HarperCollinsPublishers
77–85 Fulham Palace Road
Hammersmith
London
W6 8JB

Browse the complete Collins catalogue at
www.collinseducation.com

10 9 8 7 6 5 4 3 2

ISBN 978-0-00-741607-3

British Library Cataloguing in Publication Data
A Catalogue record for this publication is available from the British Library.

Project managed by Hart McLeod Limited, Cambridge

Edited, proofread, indexed and designed by
Hart McLeod Limited, Cambridge

Printed and bound in China

Acknowledgements

The Authors and Publishers are grateful to the following for permission to reproduce photographs.
p17 ©Craig Smith/istock.com

Whilst every effort has been made to trace the copyright holders, in cases where this has been unsuccessful, or if any have inadvertently been overlooked, the Publishers would gladly receive any information enabling them to rectify any error or omission at the first opportunity.

About this book

This book covers the content you will need to revise for GCSE Chemistry AQA A Higher. It is designed to help you get the best grade in your GCSE Chemistry Higher Exam.

The content exactly matches the topics you will be studying for your examinations. The book is divided into two major parts: **Revision guide** and **Workbook**.

Begin by revising a topic in the Revision guide section, then test yourself by answering the exam-style questions for that topic in the Workbook section.

Workbook answers are provided in a detachable section at the end of the book.

Revision guide

The Revision guide (pages 6–51) summarises the content of the exam specification and acts as a memory jogger. The material is divided into grades. There is a question (**Improve your grade**) on each page that will help you to check your progress. Typical answers to these questions and examiner's comments, are provided at the end of the Revision guide section (pages 52–57) for you to compare with your responses. This will help you to improve your answers in the future.

At the end of each module, you will find a **Summary** page. This highlights some important facts from each module.

Workbook

The Workbook (pages 65–113) allows you to work at your own pace on some typical exam-style questions. You will find that the actual GCSE questions are more likely to test knowledge and understanding across topics. However, the aim of the Revision guide and Workbook is to guide you through each topic so that you can identify your areas of strength and weakness.

The Workbook also contains example questions that require longer answers (**Extended response questions**). You will find one question that is similar to these in each section of your written exam papers. The quality of your written communication will be assessed when you answer these questions in the exam, so practise writing longer answers, using sentences. The **Answers** to all the questions in the Workbook are detachable for flexible practice and can be found on pages 121–134.

At the end of the Workbook there is a series of **Revision checklists** that you can use to tick off the topics when you are confident about them and understand certain key ideas.

Additional features

Throughout the Revision Guide there are **Exam tips** to give additional exam advice, **Remember boxes** pick out key facts and a series of **How Science Works** features, all to aid your revision.

The **Glossary** allows quick reference to the definitions of scientific terms covered in the Revision guide.

Contents

C3 Chemistry

Atoms, elements and compounds

Composition and structure

D–C

- Mixtures can be separated into simpler substances with fewer parts.
- A compound can be broken down into simpler compounds or its elements.
- Elements are substances that cannot be broken down by chemical reactions into simpler substances.

Modelling structures

B–A*

- Symbols are not always the letters of the name of the element. Sodium has the symbol Na from Natrium, the Latin word for salt.
- Chemists use models to represent the 3D arrangement of atoms. Coloured spheres represent atoms.

Inside the atom

Sub-atomic particles

D–C

- Sub-atomic particles have both a charge and a mass. You can work out the mass of an atom by adding together the numbers of protons and neutrons.
- Since an atom has equal numbers of protons and electrons, the charges cancel out.
- Neutrons have no charge and are neutral.
- The sum of the number of protons and neutrons in an atom is its mass number.

How are the electrons arranged?

- The electronic configuration of an atom gives the number of its electrons and the arrangement of the shells.
- You can show electrons as dots or crosses.
- Each shell can hold a limited number of electrons.
- The first shell (lowest energy level) can hold up to 2 electrons.
- The second shell (next energy level) can hold up to 8 electrons.
- The third shell (third energy level) can hold up to 18 electrons, but fills up with only eight before the fourth shell is started.

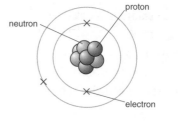

particle	relative mass	charge
proton	1	+
neutron	1	neutral
electron	almost 0	–

Figure 1: The structure of a lithium atom and sub-atomic particle chart

Remember!
In an atom the number of protons and electrons is always identical so the charges cancel out.

Why is electronic configuration so important?

B–A*

- Elements with the same number of outer electrons have similar chemical properties.

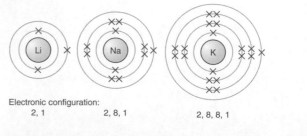

Electronic configuration:
2, 1   2, 8, 1   2, 8, 8, 1

Figure 2: Electronic configuration

Improve your grade

Draw the electronic structures of the following atoms with proton numbers 3, 9, 11, 16, and 20.
AO2 (5 marks)

Element patterns

Reactive and unreactive

- The periodic table (see page 117) lists all the known elements by increasing proton number. It has horizontal rows called Periods, and vertical columns called Groups.
- The Group number tells you how many outer electrons the atom has.
- The number of outer electrons in the atom determines how an element reacts.

D–C

Why are the elements arranged in this pattern?

- The periodic table lists the elements in order of atomic number. Noble gases (Group 0) are unreactive because they have full outer electron shells.
- After each Noble gas, the next period (row) on the table starts.

B–A*

Combining atoms

Eight in a shell

- Atoms with fewer than eight outer electrons react in ways that give them a stable group of eight in their outer shell.

- In a water molecule (H_2O), an oxygen atom needs two more electrons to have eight in its outer shell. A hydrogen atom needs one more to become stable (two electrons in the first shell). Oxygen forms two covalent bonds, one with each of two hydrogen atoms. This way all the atoms achieve a noble gas configuration.

- To form an ionic bond in sodium chloride the chlorine atom gains an electron to get an outer shell of eight. It is now a negatively charged chloride ion, Cl^-. The electron comes from the sodium atom, leaving the second shell of the sodium atom as its outer shell. This has eight electrons, so it is stable. The opposite charges attract.

Figure 3: Covalent bonding in water molecules

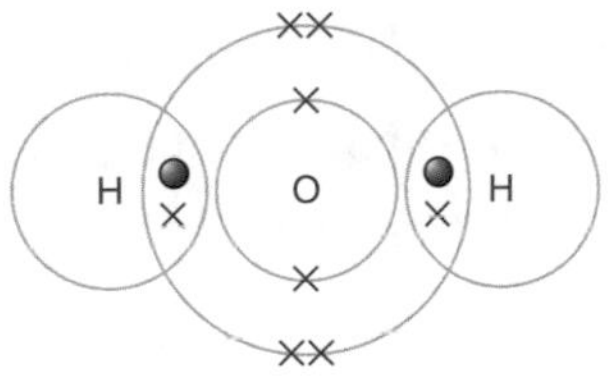

Figure 4: Ionic bonding in sodium chloride

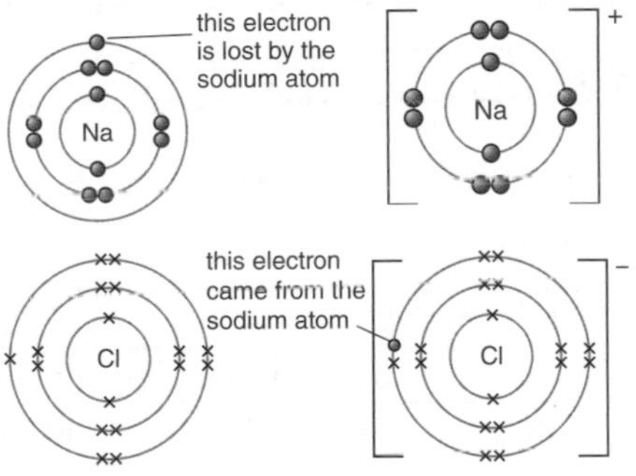

this electron is lost by the sodium atom

this electron came from the sodium atom

Remember!
Atoms with fewer than eight outer electrons react in ways that give them a stable group of eight in their outer shell. They may share electrons (covalent bonding), or transfer electrons (ionic bonding).

D–C

Representing covalent bonds

- Bonds are often represented as a short line between element symbols, for example F–F represents a fluorine molecule, a single bond joins the elements together.
- O=O represents an oxygen molecule where the two atoms are joined together by a covalent double bond.

EXAM TIP

Remember to use your periodic table to check how many outer electrons an atom has when answering questions on atomic structure and bonding. You do not need to learn this information.

B–A*

Improve your grade

Explain why oxygen has a molecule with a double covalent bond and fluorine has only a single covalent bond. You should refer to the number of outer electrons in both atoms in your answer. AO2 (4 marks)

Chemical equations

Tracking atoms and molecules

D–C

- When methane burns, the atoms in methane (CH_4) and oxygen (O_2) react and turn into carbon dioxide (CO_2) and water (H_2O).

- Figure 1 shows the molecules, but doesn't account for all the reactant atoms as products. Figure 2 shows what happens to all the atoms, it is balanced.

- This is the balanced symbol equation;
$CH_4 + 2O_2 \rightarrow CO_2 + 2H_2O$

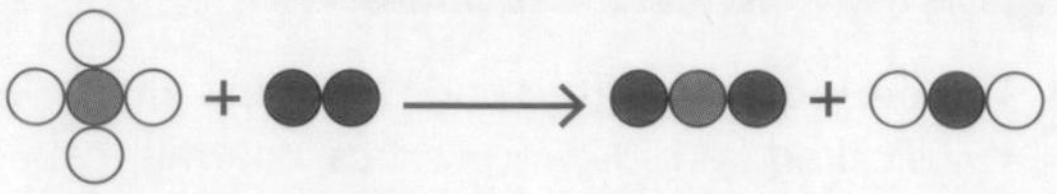

Figure 1: Methane burning with oxygen to form carbon dioxide and water

Figure 2: Methane burning with oxygen showing what happens to all the atoms

Balancing equations

B–A*

- To make an equation balance, you need to make sure you have the same number of each element on each side of the equation. To do this you add whole numbers in front of formulae as necessary. You must not change any of the formulae, because a different formula would represent a different substance.

- To balance an equation start with the metal atoms, and make sure you have the same number of reactants as the products. If needed put a whole number before the formula containing the atom.

- Now count the other atoms and balance them, leave hydrogen and oxygen to the end.

- Finish off by counting the hydrogens, then the oxygens, and correct them too. You should now have the same number of each element on each side of the equation.

Building with limestone

What are the effects of a limestone quarry?

D–C

- Limestone is obtained by quarrying. The table shows some advantages and disadvantages of quarrying limestone.

Table 1

Advantage	Disadvantage
Jobs for local people in an area with little industry	Damage to the landscape. Loss of wildlife habitats
More, better-paid jobs, so more money to boost local economy	Noise and vibration from blasting, machinery and vehicles
Better healthcare and leisure facilities, as more people move into the area	Dust pollution in the environment
Better transport links, needed for lorries or railways	Traffic congestion and vibration from heavy lorries

Limestone caves

B–A*

- Water and carbon dioxide make carbonic acid
$H_2O(\ell) + CO_2(g) \rightarrow H_2CO_3(aq)$

- Rain containing carbonic acid dissolves limestone making a solution of **calcium** hydrogen carbonate
$H_2CO_3(aq) + CaCO_3(s) \rightarrow Ca(HCO_3)_2(aq)$

- This produces caves and potholes. When the water evaporates it reverses the process producing stalactites and stalagmites (solid calcium carbonate).

Remember!
In formulae and equations capital letters are always the first letter of a symbol. The second letter is always small. Two capital letters do not make a symbol, it's a compound!

Improve your grade

Balance these two chemical equations
AO2 (2 marks)

a $Mg(s) + O_2(g) \rightarrow MgO(s)$
b $C_3H_8(g) + O_2(g) \rightarrow CO_2(g) + H_2O(\ell)$

Heating limestone

Cement

- Calcium hydroxide solution (limewater) is used to test for carbon dioxide. If carbon dioxide is bubbled through limewater, it turns cloudy or 'milky' as tiny solid particles of white calcium carbonate form.

- Cement is made by heating limestone and clay. Cement is used widely – on its own, and in mortar and concrete.

- Mortar is made by mixing cement, sand and water. Mortar binds bricks together in brick walls.

- Concrete is made by mixing cement, sand, gravel (or crushed rock) and water. Concrete is very strong. It is used for the foundations of buildings and for structures such as bridges.

D–C

The difference between mortar and concrete

- Cement, mortar and concrete do not dry out. The cement reacts with the water to form crystals that 'cement' the mixtures together.

- The mix of small sand particles and various sized stones in concrete makes it much stronger than mortar.

- Concrete can be made even stronger by reinforcing it with steel.

B–A*

Metals from ores

Making use of ores

- Only minerals with enough metal to make it worth extracting are used as ores.

- Some ores are metal oxides. These can be smelted directly.

- At the smelter, the ore is crushed and concentrated, to remove rock with little or no metal.

- Other ores are converted to the metal oxide before or during smelting.

- To convert the metal oxide to the metal, the oxygen must be removed. This is called reduction.

- The metal oxide is reduced by heating it in a furnace with carbon if the metal is below carbon in the reactivity series. Originally, the carbon was charcoal, but now it is coke (a nearly pure form of carbon, from coal).

- Limestone is often added, to remove impurities in the ore forming slag.

Table 2: The reactivity series

potassium	K
sodium	Na
calcium	Ca
magnesium	Mg
aluminium	Al
carbon	C
zinc	Zn
iron	Fe
lead	Pb
hydrogen	H
copper	Cu
silver	Ag
gold	Au

Increasing reactivity →

D–C

Extracting more reactive metals

- Many metals were not discovered until the discovery of electricity in the 1800s.

- Metals above carbon in the reactivity series are extracted using electrolysis.

- Electrolysis involves passing an electric current through a molten metal compound, splitting it into its metal and non-metal elements.

Electricity
Molten aluminium oxide → molten aluminium + oxygen gas

Remember!
When asked about the environmental or social impact of quarrying limestone make sure you give both advantages and disadvantages if you are to get full marks.

B–A*

EXAM TIP

Always check the reactivity series before answering questions about metal extraction. Metals below carbon always use carbon to reduce them, metals above are reduced by electrolysis.

Improve your grade

Draw a table to show which of these metals is extracted by carbon and which by electrolysis. **AO1 (2 marks)**
iron, magnesium, aluminium, copper, zinc, lead, potassium, and calcium

Extracting iron

Inside a blast furnace

D–C

- Iron ore (Fe_2O_3), coke and limestone are fed in at the top of the furnace.

- Molten iron and slag collect at the bottom.

- As the hot air is blown through, the coke burns in oxygen to produce carbon monoxide:
  carbon + oxygen → carbon dioxide
  $$C(s) + O_2(g) \rightarrow CO_2(g)$$
  carbon dioxide + carbon → carbon monoxide
  $$CO_2(g) + C(s) \rightarrow 2CO(g)$$

- The carbon monoxide then reduces the iron oxide to iron, and is oxidised to carbon dioxide:
  $$Fe_2O_3(s) + 3CO(g) \rightarrow 2Fe(\ell) + 3CO_2(g)$$

- The limestone reacts with impurities in the ore to make slag, this floats on top of the iron.

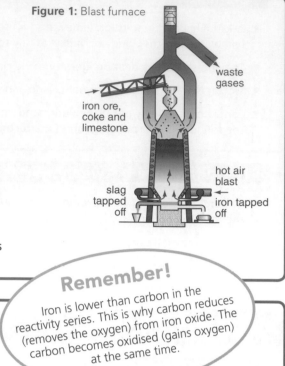

Figure 1: Blast furnace

waste gases

iron ore, coke and limestone

hot air blast

slag tapped off

iron tapped off

Oxidation and reduction

B–A*

- Oxidation reactions occur when oxygen is added to a substance.

- Reduction reactions occur when oxygen is removed from a substance.

- Oxidation is the reverse of reduction and both must always occur together.

Remember!

Iron is lower than carbon in the reactivity series. This is why carbon reduces (removes the oxygen) from iron oxide. The carbon becomes oxidised (gains oxygen) at the same time.

Metals are useful

Atoms and alloys

D–C

- Metal atoms are arranged in giant structures, in regular rows and layers.

- Metals can bend because these layers can slide over each other. The shape can change but the atoms remain bonded together.

- Metals conduct electricity because some of their outer electrons are free to move through the layers.

- To make metals harder they can be mixed to form alloys.

- An alloy is not a compound but is a mixture of elements, usually metals. Steel is an alloy of iron and carbon. Brass is an alloy of copper and zinc.

- The proportions of each element in an alloy can differ. This affects the alloy's properties and uses.

- The more carbon in steel the harder it is. Nine carat gold has more copper in it than 22 carat gold, and is harder wearing.

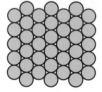

Figure 2: Layer of copper atoms

Smart alloys

B–A*

- These alloys can remember their original shape, and if warmed return to that shape.

- To mend broken bones, strips of smart alloy are cooled, stretched and screwed onto the bone. As the strips warm up they shrink, pulling the bones back together so the break heals faster and in the correct position.

Improve your grade

Explain the difference between oxidation and reduction. **AO1 (2 marks)**

Iron and steel

Designer steels

- Pure iron is too soft for most uses, so steels are used instead.

- Adding other metals to molten steel can give it special properties. The choice of metal depends on how the steel will be used, and therefore the properties required.

- Stainless steel is about 70% iron, 20% chromium and 10% nickel. Stainless steel is very resistant to corrosion and does not rust.

Table 1 How other metals change steel's properties

Metal added to steel	Improvement to steel properties
Chromium and/or nickel	More corrosion resistance
Manganese	More strength and hardness
Molybdenum and/or tungsten	More strength, hardness and toughness
Vanadium	More strength, less brittle

D–C

Why is steel harder than iron?

- The atoms in pure metals have the structure shown in Figure 2. If other different sized atoms are added to make an alloy, as in Figure 3, the layers of atoms find it hard to slide past each other and so the metal is harder.

Figure 3: Different sized atoms give steels hardness and strength

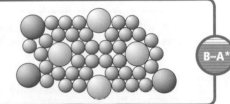

B–A*

Copper

Physical and chemical changes during extraction

- Copper is purified using electrolysis.

- Electricity is passed through a copper sulfate solution. This is called electrolysis.

- Copper atoms in the impure positive electrodes lose electrons. They become copper ions (Cu^{2+}) and dissolve into the solution. The impurities fall to the bottom.

- At the pure copper negative electrodes, copper ions in the solution gain electrons and coat the electrode with copper atoms.

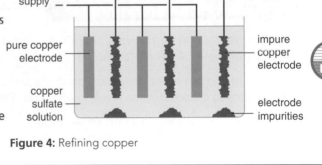

electricity supply + −

pure copper electrode

impure copper electrode

copper sulfate solution

electrode impurities

Figure 4: Refining copper

D–C

Extracting copper from low grade ores

- High quality copper ores are very limited so new ways of cheaply producing copper ores are being developed.
  - Bioleaching – Bacteria convert insoluble copper compounds into soluble ones.
  - Phytomining – Plants absorb copper compounds through their roots. After harvesting, the plants are burnt leaving a copper-rich ash.

How Science Works

- When testing properties of new materials such as alloys, the test must be both repeatable (gives the same result every time you carry out the test), and reproducible (gives the same conclusion when done by someone else, or using a different method).

B–A*

Improve your grade

Explain why the waste heaps of an old copper mine would be a suitable site for phytomining or bioleaching.
AO2 (2 marks)

Aluminium and titanium

Extraction

D–C

- Aluminium is very expensive as it can only be extracted using electricity.

- The ore bauxite (Al_2O_3) is dissolved in cryolite to allow it to melt at 900 °C.

- The molten aluminium oxide then is electrolysed:
  aluminium oxide → aluminium + oxygen
  $$2Al_2O_3 \quad \rightarrow \quad 4Al \quad + \quad 3O_2$$

- Each aluminium ion needs **three electrons** to become an atom. This is why so much electricity is needed.
  $$Al^{3+} + 3e^- \rightarrow Al$$

- Titanium cannot be extracted by carbon or electricity. First the ore, rutile (titanium dioxide) is converted to titanium chloride. Then titanium chloride is reacted with magnesium.

- Titanium chloride + magnesium → titanium + magnesium chloride.
  $$TiCl_4 \quad + \quad 2Mg \quad \rightarrow \quad Ti \quad + \quad 2MgCl_2$$
  This method is very costly which means that titanium is also very expensive.

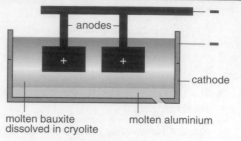

Figure 1: Extracting aluminium by electrolysis

Remember!
Using cryolite saves money as bauxite normally melts at 2000 °C. Reducing the temperature saves energy and money.

Corrosion resistance

B–A*

- Both titanium and aluminium are very reactive metals, yet they resist corrosion.

- Both metals react easily with oxygen forming a tough oxide layer on the surface, preventing further reaction.

- Acids and alkalis do attack aluminium, because they react with the oxide layer. Titanium oxide does not react so titanium can be used inside our bodies for replacement joints and to hold broken bones together.

Metals and the environment

Effects on the environment

D–C

- Mining metal ores destroys the landscape, wildlife habitats, and displaces the local people, changing their way of life.

- Using carbon to reduce metal ores produces carbon dioxide. This adds to carbon dioxide levels in the atmosphere.

- Smelting ores containing sulfur produces sulfur dioxide which causes acid rain. If collected, the sulfur dioxide can be used to make useful sulfuric acid.

- Recycling aluminium and steel saves energy and the environment. Aluminium cans are cleaned and melted before reusing the metal. Iron and steel are added as scrap to steel-making furnaces.

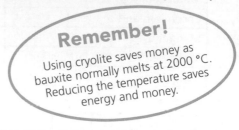

Figure 2: How much waste rock is mined, and later dumped, for each 1000 kg copper produced?

Recovering metals from brownfield sites

B–A*

- Brownfield sites are areas of land that have been used before. They are often polluted by toxic metal compounds of cadmium, nickel and cobalt.

- Instead of removing the soil from these sites, developers now grow plants on them. These plants absorb the toxic metal ions making the land safe to use.

- When burnt the plant ash contains compounds of the toxic metals that can be used as ores. This is phytomining.

Remember!
Phytomining takes time, but it is cheap, and the metal compounds are easily obtained at the end.

Improve your grade

Explain why titanium and aluminium metals are both reactive and corrosion free. **AO1 (2 marks)**

A burning problem

Global effects

- Global warming. Increasing carbon dioxide in the atmosphere traps more energy from the Sun making the Earth warm up.

- Global dimming. Dust particles in the air prevent sunlight reaching the ground. This reduces the energy available for photosynthesis, and may cool the Earth as well.

- Both may affect the weather patterns on the Earth.

- Acid rain is caused by nitrogen, sulfur and carbon oxides dissolving in rain water. The acid rain formed attacks limestone buildings much more quickly than normal rain. It can also cause serious damage to trees and to aquatic life in affected lakes.

D–C

Other problems with combustion

- During combustion, each carbon atom in a fuel needs two oxygen atoms to form carbon dioxide, CO_2.

- If there is not enough air, some carbon atoms get only one oxygen atom (carbon monoxide), or none at all (carbon or soot).

- Carbon monoxide is poisonous, it is easily absorbed by your red blood cells instead of oxygen, therefore depriving your body of oxygen.

B–A*

Reducing air pollution

Alternative solutions

- Vehicles that burn fossil fuels produce poisonous carbon monoxide, and they also produce nitrogen oxides which can cause acid rain.

- To reduce the amount of these compounds released by vehicles all new cars have catalytic converters fitted.

- Carbon monoxide is oxidised to carbon dioxide, and nitrogen oxides are reduced to nitrogen.

- Alternative fuels are also being used, such as biodiesel made from vegetable oils, which contain almost no sulfur. Ethanol produced by fermentation from sugar, and hydrogen, are sulfur free.

D–C

Problems with alternatives

- Rainforest is cut down to grow sugar and soya bean for biofuels. Elsewhere, land that could grow crops to feed the world's increasing population is, instead, producing fuels. This forces food prices up for every one.

- Hydrogen can only be obtained by using electricity to electrolyse water. Electric cars are emission free. However, producing electricity does produce harmful pollution, and whilst the vehicles do not produce pollution, making the electricity they need does.

B–A*

EXAM TIP

When asked to evaluate the benefits and risks make sure you consider both the benefits and also the drawbacks in your answer. Be specific, 'food prices will rise' is better than 'things will cost more', 'less sulfur oxides will be produced' is better than 'pollution will be less'.

⊙ Improve your grade

Suggest why using biofuels may create more problems than it solves. **AO2 (3 marks)**

Crude oil

D–C

How does fractional distillation work?

- When crude oil is heated to about 400 °C, most of it boils and vaporises.

- Vapours consisting of hydrocarbons rise up the column, gradually cooling. When cooled below their boiling point, they condense back to liquid and are run off.

- Hydrocarbons with high boiling points condense first. The lower their boiling point, the higher up the column they rise before condensing.

- Hydrocarbons with different size molecules condense at different levels, separating the crude oil mixture into a series of fractions with similar numbers of carbon atoms and boiling points.

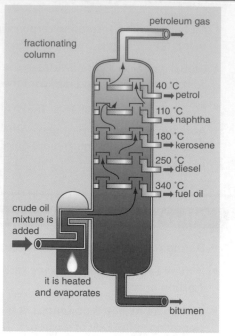

Figure 1: Fractional distillation of oil

Name of fraction	Carbon atoms per molecule	Uses
petroleum gas	1 to 4	heating, cooking, LPG fuel
petrol	5 to 9	fuel (cars and lorries)
naphtha	6 to 10	to make other chemicals
kerosene	10 to 16	jet fuel, paraffin
diesel	14 to 20	fuel (cars, lorries and trains)
fuel oil	20 to 50	fuel for ships, factories and heating
bitumen	more than 50	tar for road making

Why do boiling points depend on size?

B–A*

- There are strong covalent bonds between the atoms in hydrocarbon molecules.

- There are only weak attractions between the hydrocarbon molecules. With small molecules there are fewer attractions between the molecules.

- Bigger molecules have more bonds between them, and more attractions that need to be broken to make the molecules separate into a gas, so they need more energy (a higher temperature).

Alkanes

Patterns and properties

D–C

- The general formula for alkanes is C_nH_{2n+2} where n is the number of carbon atoms.

- All alkanes have similar chemical properties.

- As the number of carbon atoms increases;
  - molecules become larger and heavier
  - boiling point increases
  - flammability decreases (catch fire less easily)
  - viscosity increases (liquid becomes thicker).

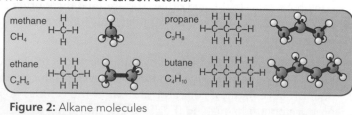

Figure 2: Alkane molecules

What makes alkanes similar to each other?

B–A*

- Alkanes only have single covalent bonds joining each atom together. They are said to be saturated hydrocarbons.

- Alkanes have similar structures so they react in similar ways.

EXAM TIP

When drawing molecules of hydrocarbons, remember each carbon atom needs four bonds, and each hydrogen atom needs just one bond.

Improve your grade

Draw out the structure of the straight chained alkanes with 5 carbon atoms and 7 carbon atoms. **AO2 (2 marks)**

Cracking

How cracking works

D–C

- Crude oils have different amounts of different hydrocarbons, but never enough petrol sized molecules for our needs.

- Heating the hydrocarbons makes each molecule waggle until a carbon–carbon bond breaks. They break in different places, giving a mixture of products. So $C_{10}H_{22}$ could crack to form;
  $C_8H_{18} + C_2H_4$ or $C_7H_{16} + C_3H_6$ or $C_6H_{14} + C_4H_8$ and so on.

Figure 3: Cracking decane

- One of the molecules made is an alkane, the other has two bonds (a double bond) shown as C=C between a pair of carbon atoms, and is called an alkene.

- Alkenes have the general formula C_nH_{2n}

- The double bond in alkenes makes them much more **reactive** than alkanes. This makes alkenes extremely useful.

- Ethene is a particularly important alkene product of cracking. It is the starting point for making polythene and many other plastics.

Remember!
There is no alkene with only 1 carbon atom. They need at least two carbon atoms. Each carbon atom still needs to have four bonds, the double bond counts as two!

Types of cracking

B–A*

- Fuel oil is mixed with steam in a furnace at about 850 °C. The hydrocarbons undergo thermal decomposition. Changing the amount of steam alters the products made.

- Fuel oil is vaporised and mixed with a catalyst at about 600 °C. Using a catalyst allows the cracking reaction to take place at a lower temperature than in steam cracking.

Alkenes

Reactive alkenes

D–C

- A double bond is just two bonds.

- One of the two bonds can 'open up', allowing each carbon atom to form a bond with another atom. This means that alkenes are reactive compounds.

- Each carbon atom in an alkane already has bonds to four other atoms. So, unlike alkenes, alkanes cannot react by adding extra atoms. Alkanes are saturated – they cannot add any more atoms. Alkenes can, so alkenes are unsaturated.

- Fats are more complex than hydrocarbons, but we also refer to them as being saturated and unsaturated. Polyunsaturated fats have lots of double bonds.

Detecting double bonds

B–A*

- Adding a few drops of orange-brown bromine water to a sample of a hydrocarbon shows whether the hydrocarbon is saturated.

- If the hydrocarbon contains double bonds, the orange brown colour will rapidly disappear, it is unsaturated. If the bromine water stays orange-brown, then the hydrocarbon is saturated.

Remember!
When your clothes are so wet they cannot accept any more water we say they are saturated. Hydrocarbons that cannot accept any more atoms, because all the bonds are single, are also called saturated.

Improve your grade

Draw structural diagrams to show how $C_{12}H_{26}$ can be converted to C_3H_6, and another molecule. State which of the two products is unsaturated, and why. **AO1 and 2 (3 marks)**

Making ethanol

Converting into ethanol

D–C

- Alcohol is the name of another chemical family.

- Alcohols are compounds with a hydroxyl group (–OH).

- Ethanol is a member of the alcohol family, ethanol has two carbons. Its formula is C_2H_5OH– written like that, rather than C_2H_6O, to show that it has the –OH group.
  - Ethanol is made by two methods:
  - from crude oil. Cracking large alkanes produces ethene. Blowing ethene and steam over a hot catalyst makes ethanol.
  - from sugar. Adding yeast to sugar dissolved in water causes fermentation and changes sugar into ethanol.

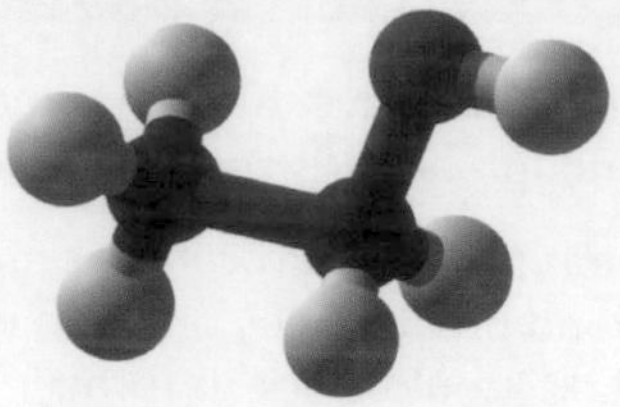

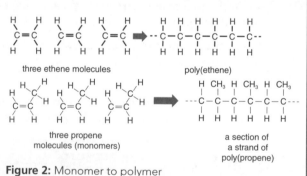

Figure 1: Molecular model of ethanol

- Both methods give a solution of ethanol in water. Ethanol is separated by fractional distillation.

- Fermentation happens when micro-organisms, called yeasts, feed on the sugar and convert it into ethanol.

- To manufacture ethanol for use as a fuel, a sugar solution made from sugar cane or maize (corn) is prepared. Yeast is added to cause fermentation.   sugar → ethanol + carbon dioxide

- Fuels obtained from animals and plants are called biofuels, so ethanol made by fermentation is a biofuel.

- As more plants can be grown after harvesting, biofuels are a renewable source of energy. Unfortunately it uses land and crops that could be used to feed people.

Other uses of ethanol

B–A*

- Some substances do not dissolve in water, but do dissolve in ethanol.

- This is useful when making pharmaceuticals (medicines), perfumes and aftershaves, inks and varnishes.

- It is the major ingredient in surgical spirit (an antiseptic that kills pathogens).

- It is used to make a wide range of other chemicals, including flavourings and perfumes.

Polymers from alkenes

A variety of polymers

D–C

- To make poly(ethene), ethene is heated under pressure. Ethene is known as the monomer. A catalyst sets off a chain reaction. It makes a C=C bond open up and join onto another ethene molecule. That double bond then opens up and joins onto the next, and so on, forming a polymer chain.

- The reaction involves only the C=C double bond. It does not matter what other atoms are attached to the carbons. By changing the other atoms attached to the carbons, it is possible to produce lots of other different polymers with different properties.

three ethene molecules → poly(ethene)

three propene molecules (monomers) → a section of a strand of poly(propene)

Figure 2: Monomer to polymer

Polymers and plastics

B–A*

- Polymers are molecules consisting of very long chains made from carbon atoms, with various other atoms attached. Many natural materials are polymers, including silk, rubber, starch, proteins and DNA.

- Most polymers have low densities, are not brittle (and do not break easily), can withstand corrosion by chemicals and soften when heated (can easily be moulded into shape).

Improve your grade

PVC (polyvinylchloride) is a commonly used polymer. Its monomer has the structure:
Show, using three monomer molecules, the structure and bonding of a PVC polymer chain. **AO2 (2 marks)**

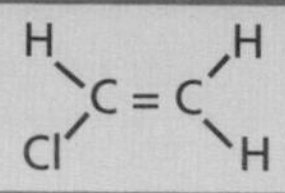

Designer polymers

Special polymers

- Plastic polymers are easily moulded into shape, low density (lightweight), waterproof and resistant to acids and alkalis.

- Properties such as strength, hardness and flexibility vary.

- Polymers can be designed to have the specific properties needed for a particular purpose.

D–C

What about the future?

Smart polymer materials can, for example, respond to changes such as temperature or voltage.

- Conducting polymers can conduct electricity.

- Light-emitting polymers give off light when electricity passes through.

- Shape memory polymers 'remember' the shape of the object.

- Biodegradable polymers rot away, unlike other polymers that cause waste problems.

B–A*

Polymers and waste

Sorting out waste

- Most disposable plastic items are labelled with a recycling symbol and code number to identify the polymer.

- Instead of dumping in landfills, household rubbish, including plastics, can be burned in incinerators. The heat produced may be used to generate electricity or heat local buildings. This is wasteful of polymers and can cause air pollution.

D–C

Biodegradable plastics

- Plastics in landfill do not rot, and the ground will be unusable for agriculture.

- A few plastics are water-soluble and/or biodegradable:
  - Biodegradable plastic carrier bags are made from polythene and corn starch. The cornstarch is decomposed by micro-organisms leaving the polythene in microscopic pieces.
  - Biopol® is a biodegradable plastic produced by micro-organisms. It is used in medicine to make stitches and hold bones together while they heal. It dissolves away so you don't need to have the stitches removed.

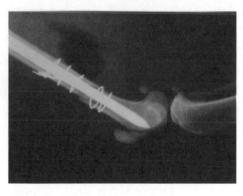

Figure 3: Polymer pins used to fix a broken bone

B–A*

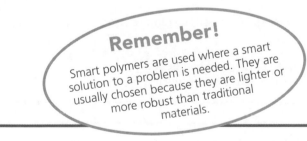

Remember!

Smart polymers are used where a smart solution to a problem is needed. They are usually chosen because they are lighter or more robust than traditional materials.

Improve your grade

Describe the difference between recycling polymer waste and re-using it by incineration. Give two reasons why recycling is more environmentally friendly than incineration. **AO1 (3 marks)**

Oils from plants

Oils in food and fuel

- Plant oils store a lot of energy. Oils provide more energy than most other foods.

- Oils from rapeseed, soya beans and other crops are converted into biodiesel fuel.

- Cooking food in oils produces different flavours and textures; the food is cooked faster at a higher temperature than in water. The food absorbs some oil increasing its energy content.

- Oils also contain other nutrients we need.
  - Essential fatty acids (for example omega 3 and 6), for the heart, muscles and nervous system to function properly.
  - Vitamins. Seed and nut oils are particularly rich in vitamin E.
  - Minerals. Minerals are compounds of metals and non-metals such as potassium, calcium, iron and phosphorus.
  - Trace elements, but only in tiny amounts. For instance, a single Brazil nut provides the whole recommended daily allowance (RDA) of selenium.

Essential oils

- Flowers contain essential oils, and are different from the natural oils in seeds and nuts. They have low boiling points, so evaporate easily, giving flowers their scents. They are used in perfumes.

- Essential oils are extracted using steam distillation. The less dense essential oil floats on top of the water.

- Vegetable oils have much more complex molecules than mineral oil, with many carbon, hydrogen and oxygen atoms.

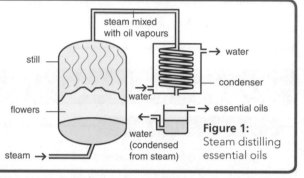

Figure 1: Steam distilling essential oils

Biofuels

Biofuel issues

Biofuels are made from animal or plant material, for example wood.

- Biofuels contain no sulfur and this advantage means that they do not cause sulfur dioxide pollution.

- Biofuels may be renewable, but they have problems of their own.
  Practical issues:
  - plant oils are very viscous so are hard to use directly in car engines
  - they need more air to burn properly than diesel and petrol
  - few garages sell biofuels as not many people currently use them.
  - Economic and ethical issues:
  - biofuels produce less energy per litre, so a greater volume is needed
  - growing crops needs energy for the machinery, fertilisers and transport, this may be more than is produced by the biofuel crop
  - land used to grow food crops may be used to grow biofuels instead, leading to food shortages and raised food prices.
  Environmental issues:
  - the demand for fuel is so great that huge amounts of land would be needed
  - changing land use can affect the plants and animals that live there, reducing biodiversity.

> **EXAM TIP**
>
> If asked to evaluate the impact of biofuels on the environment or another issue. Give both benefits and problems in your answer to get full marks.

Making biodiesel

- Plant oils, animal fats or used cooking oil can be converted to biodiesel and then used in diesel engines.

- Reacting the oils or fats with methanol converts them to biodiesel, which is a mixture of chemicals called esters.

Improve your grade

Explain why biofuels are considered to be carbon neutral. **AO1 (2 marks)**

Oils and fats

Unsaturated fats

- Testing an unsaturated fat is similar to testing for an alkene. They both contain double C=C bonds.
- The amount of bromine water decolorised by the fat indicates how many C=C bonds there are in the fat.
- Everyone needs some fats in their diet, for energy and essential nutrients, but too much fat is unhealthy.
- Saturated fats can increase the level of cholesterol in the blood. You should;
  - eat foods rich in polyunsaturates, such as sunflower oil, and monounsaturates, such as olive oil
  - avoid saturated fats, such as lard, limit the amount of fat of all types in your food,
  - eat fish, some fish oils contain omega-3 fatty acids. These have been shown to lower blood pressure.

D–C

Hardening vegetable oils

- Liquid vegetable oils can be hardened into solid spreads by hydrogenation.
- Hydrogenation adds hydrogen to the double bonds so the oil becomes saturated and solid.
- Hydrogenation uses a nickel catalyst and a temperature of about 150 °C.

B–A*

unsaturated oil (liquid)   saturated fat (solid)

Figure 2: Hydrogenation of a C=C double bond

Emulsions

Useful emulsions

- Oil and water are immiscible. However they can be made to mix by stirring or shaking, forming an emulsion.
- Emulsions are less viscous (less runny) than the oil and more viscous than the aqueous solution.
- Oil-in-water emulsions contain droplets of oil suspended in an aqueous solution.

D–C

Emulsifiers

- Many emulsions are unstable and rapidly separate back to oil and water.
- Emulsifiers are compounds whose molecules have opposite properties at each end.
  - One end is hydrophilic (water-loving) – it is attracted to water but not to oil.
  - The other end is hydrophobic (water-fearing) – it gets away from the water by sticking into an oil droplet.
- The oil droplets become surrounded by emulsifier molecules with their hydrophobic ends in the oil. The hydrophilic ends on the outside attract the droplets to the water. This also prevents the oil droplets joining together to form a separate oil layer again.

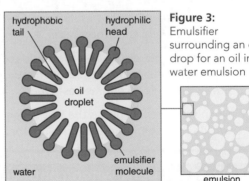

hydrophobic tail   hydrophilic head

oil droplet

water   emulsifier molecule

emulsion

Figure 3: Emulsifier surrounding an oil drop for an oil in water emulsion

B–A*

How Science Works

- When deciding how much emulsifier to use to make a stable emulsion it is important to control all the other variables. You should always state how to control the variables, giving examples.

Remember!
Immiscible liquids do not mix together however hard you shake them. They simply separate into two layers, the less dense layer is the top layer, which is usually an oil, and the denser water layer is below.

Improve your grade

Draw a diagram to show how an emulsifier would surround a water droplet to make a water in oil emulsion.
AO2 (2 marks)

Earth

What is Earth like inside?

D–C

- The crust is a thin outer layer of cold, solid rock. Its thickness is between 5 and 30 kilometres.

- The mantle is made of very hot molten rock that flows very slowly by convection currents.

- The core is at the centre; an inner core, in the centre, is solid iron and nickel and an outer core is a molten mixture of iron and nickel.

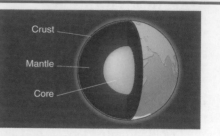

Figure 1: The layers of the Earth

How did the Earth's surface features form?

B–A*

- Earth's crust is broken up into separate parts called tectonic plates. These float on the mantle.

- Where plates push against each other, they force the land upwards, forming mountain ranges.

- In the same way, rocks such as limestone, which formed under the sea, were pushed up and became hills.

- Over millions of years, rivers gradually wear away rocks, forming valleys and spectacular canyons.

- Islands such as Iceland and Hawaii are formed when volcanoes on the sea bed throw out material which builds up.

Continents on the move

Continental drift

D–C

- Alfred Wegener in 1915 suggested that all the continents were once joined together in a super continent he called Pangea.

- When Wegener suggested his hypothesis, other scientists rejected it because no one could explain how huge continents could move. Later other scientists found evidence to support Wegener's theory.

- Earth's crust and the semi-solid upper part of the mantle make up the lithosphere.

200 million years ago   100 million years ago   50 million years ago

Figure 2: Pangea breaking up

- The tectonic plates float on the liquid rock of the mantle. Heat from radioactive processes within the Earth drives convection currents in the mantle. These currents carry the floating plates, and the continents very slowly, only a few centimetres a year, in a process called continental drift.

- When two tectonic plates the size of continents collide they push the land upwards, forming mountains. The world's largest mountain ranges are formed where continental plates collide.

What happens when continents move apart?

B–A*

- When tectonic plates move apart, magma (molten rock) escapes from the mantle, forming new crust.

- In the Atlantic Ocean as the plates move apart, magma rises through the gap creating an underwater mountain range called the Mid-Atlantic Ridge.

- Iceland is on the Mid-Atlantic ridge, where the ridge has become so high that it is no longer underwater. The two plates are still moving, so Iceland is getting wider each year.

> **EXAM TIP**
>
> Evidence for continents having been joined together includes: the shape of the continents; Africa and South America are like jigsaw pieces that can fit together; similar rock formations where the continents would have joined together; and similar fossils in rocks on both sides of the Atlantic.

Improve your grade

Explain how the Atlantic Ocean is getting wider each year. Suggest another part of the world where a similar process may be taking place. **AO1 (3 marks)**

Earthquakes and volcanoes

Volcanoes

- Volcanoes often occur at plate boundaries.
- Magma stays sealed under the crust, often for hundreds of years. Eventually the pressure builds up enough for magma to burst through a vent – a crack or weak spot in the crust. The blast creates a crater, lava and gas pour out, and the familiar cone shape of volcanoes forms.
- Scientists can monitor the movement of tectonic plates. They can tell when pressure is building and an earthquake is possible. However, they cannot obtain hard data about the forces involved, the friction between plates and structural weaknesses in plates. Eruptions are not the same as earthquakes, but the problems of predicting them are: the lack of sufficient valid data.

D–C

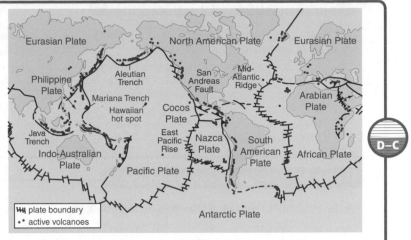

Figure 3: Main tectonic plate boundaries and active volcanoes

Subduction zones

- Where plates meet, the lighter plate, usually the oceanic one is pushed under the heavier continental plate. This is called subduction. As the lighter plate pushes into the mantle it forces magma to the surface through weaknesses in the crust (see Figure 3).

B–A*

The air we breathe

How did the atmosphere evolve?

- The Earth was formed about 4 600 million years ago. For the first 1 000 million years there was intense volcanic activity, releasing huge amounts of steam, carbon dioxide, and some ammonia (NH_3) and methane (CH_4).
- The steam condensed and eventually formed the seas and oceans.
- About 3 400 million years ago simple life that could photosynthesise had developed in the oceans and seas. These algae used the carbon dioxide and water and released oxygen. The oxygen reacted with the ammonia to make nitrogen gas.
- About 400 million years ago the atmosphere had enough oxygen to allow land plants and then animals to evolve.
- The atmosphere has stayed the same for about the last 200 million years with 78% nitrogen, 21% oxygen, and small amounts of other gases such as argon and carbon dioxide.

D–C

Separating gases from the air

- Air is dried and filtered to remove water vapour and dust.
- Air is cooled to about −200 °C. Since this is below the boiling points of nitrogen (−196 °C) and oxygen (−183 °C), the air liquefies.
- As the liquid air warms, the nitrogen boils off first, leaving liquid oxygen behind.

Figure 4: Fractional distillation of liquid air

nitrogen gas out →
−190 °C
liquefied air in at −200 °C →
−185 °C
liquid → oxygen out

B–A*

How Science Works

- You should be able to explain why scientists cannot predict accurately when earthquakes and volcanic eruptions will take place.

Improve your grade

Sketch a timeline showing how the proportions of water vapour, carbon dioxide, oxygen, and nitrogen have changed since the Earth was formed. **AO2 (4 marks)**

The atmosphere and life

How did life on Earth begin?

D–C

- There is uncertainty about how life began because there is no evidence. The first primitive life-forms did not form fossils. Here is one theory.
  - For the first billion years, Earth's atmosphere was mainly carbon dioxide, with some methane, ammonia, hydrogen and water vapour. The water vapour eventually condensed to form oceans.
  - The weather was more extreme than it is today. Frequent lightning provided energy to break chemical bonds and split molecules. The fragments recombined in different ways, forming new compounds.
  - These new compounds included amino acids (from which all proteins are built up), sugars and other carbon compounds needed to make DNA (deoxyribonucleic acid). These compounds are the basis of life.

The Miller–Urey experiment

B–A*

- In 1952 Miller and Urey, devised an experiment to test the theory of how life began. They mixed methane, ammonia and hydrogen in a sterile glass bulb. This simulated the early atmosphere.

- A flask of water represented the ocean. They circulated water vapour from the 'ocean' through the 'atmosphere', and made electric sparks to simulate lightning.

- After a week they analysed the mixture. They found many different organic compounds – carbon compounds that normally come from living organisms. These included glycine and several other amino acids, and sugars, including ribose. This was significant because D in DNA stands for deoxyribose, a closely related sugar.

- Miller and Urey did not create life, they showed that molecules essential for living things could be made by a natural process from Earth's early atmosphere.

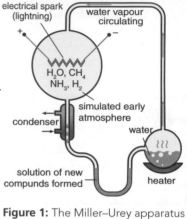

Figure 1: The Miller–Urey apparatus

Carbon dioxide levels

Carbon recycling

D–C

- Both animals and plants take in oxygen and give out carbon dioxide. (see Figure 2)

- Dead plants and animals decay. Oxygen from the air or water converts them back into carbon dioxide, with the help of bacteria, fungi, and other organisms.

- Most of the carbon dioxide in the early atmosphere became locked up as fossil fuels or in rocks such as limestone made from animal shells.

- Carbon dioxide is also absorbed by the oceans, removing it from the atmosphere.

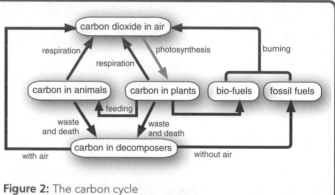

Figure 2: The carbon cycle

So what's the problem?

B–A*

- People are using more and more fossil fuels for energy. This is releasing far more carbon dioxide into the atmosphere than photosynthesis can remove.

- Carbon dioxide is a greenhouse gas. It traps energy from the Sun. Increasing amounts of carbon dioxide are likely to make the Earth hotter in future.

- Higher concentrations of carbon dioxide in the atmosphere mean that more dissolves in the seas and oceans. They become slightly more acidic (lower pH), upsetting ecosystems, killing coral for instance.

Improve your grade

Explain why the Miller–Urey experiment does not prove how life evolved. **AO3 (2 marks)**

C1 Summary

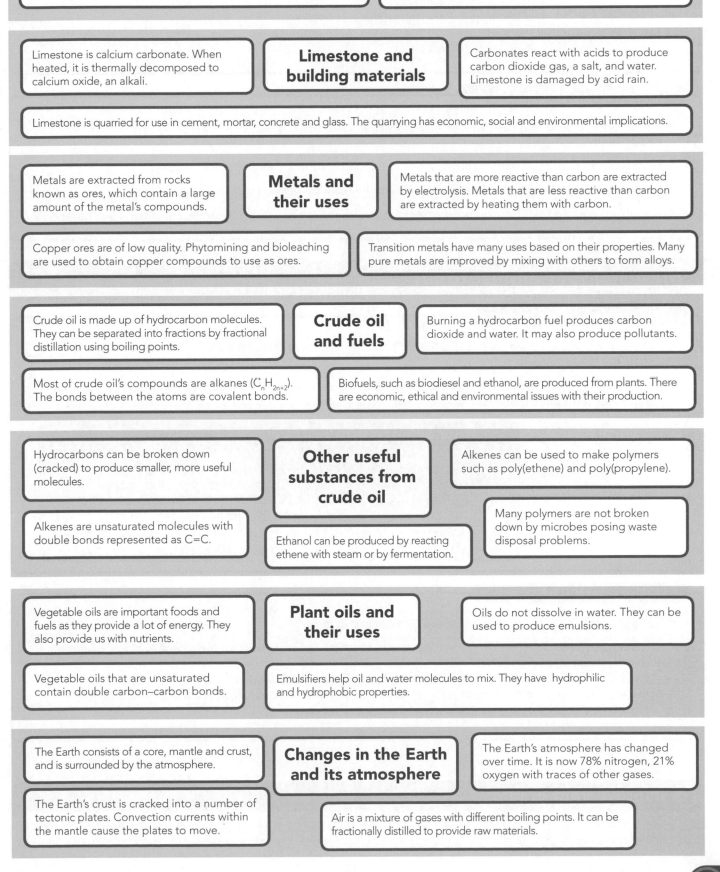

All substances are made from atoms, which have protons and neutrons in a central nucleus, with electrons in shells around the nucleus.

Fundamental ideas

Electrons occupy specific energy levels or shells but always the lowest available level. Atoms with full outer electron shells (Group 0) are stable.

The atomic number = the number of protons in the nucleus. It is the same as the number of electrons in the atom. The mass number of an element = the number of protons + neutrons.

Metals within non-metal compounds have ions held together by ionic bonds. Non-metal compounds consist of molecules with covalent bonds.

Limestone is calcium carbonate. When heated, it is thermally decomposed to calcium oxide, an alkali.

Limestone and building materials

Carbonates react with acids to produce carbon dioxide gas, a salt, and water. Limestone is damaged by acid rain.

Limestone is quarried for use in cement, mortar, concrete and glass. The quarrying has economic, social and environmental implications.

Metals are extracted from rocks known as ores, which contain a large amount of the metal's compounds.

Metals and their uses

Metals that are more reactive than carbon are extracted by electrolysis. Metals that are less reactive than carbon are extracted by heating them with carbon.

Copper ores are of low quality. Phytomining and bioleaching are used to obtain copper compounds to use as ores.

Transition metals have many uses based on their properties. Many pure metals are improved by mixing with others to form alloys.

Crude oil is made up of hydrocarbon molecules. They can be separated into fractions by fractional distillation using boiling points.

Crude oil and fuels

Burning a hydrocarbon fuel produces carbon dioxide and water. It may also produce pollutants.

Most of crude oil's compounds are alkanes (C_nH_{2n+2}). The bonds between the atoms are covalent bonds.

Biofuels, such as biodiesel and ethanol, are produced from plants. There are economic, ethical and environmental issues with their production.

Hydrocarbons can be broken down (cracked) to produce smaller, more useful molecules.

Other useful substances from crude oil

Alkenes can be used to make polymers such as poly(ethene) and poly(propylene).

Alkenes are unsaturated molecules with double bonds represented as C=C.

Ethanol can be produced by reacting ethene with steam or by fermentation.

Many polymers are not broken down by microbes posing waste disposal problems.

Vegetable oils are important foods and fuels as they provide a lot of energy. They also provide us with nutrients.

Plant oils and their uses

Oils do not dissolve in water. They can be used to produce emulsions.

Vegetable oils that are unsaturated contain double carbon–carbon bonds.

Emulsifiers help oil and water molecules to mix. They have hydrophilic and hydrophobic properties.

The Earth consists of a core, mantle and crust, and is surrounded by the atmosphere.

Changes in the Earth and its atmosphere

The Earth's atmosphere has changed over time. It is now 78% nitrogen, 21% oxygen with traces of other gases.

The Earth's crust is cracked into a number of tectonic plates. Convection currents within the mantle cause the plates to move.

Air is a mixture of gases with different boiling points. It can be fractionally distilled to provide raw materials.

Investigating atoms

How ideas about atoms have changed

D–C

- All matter consists of atoms, which cannot be divided up.
- All atoms of the same element are identical, but different from atoms of every other element.
- The Greeks suggested the idea of atoms in the 5th century BC, John Dalton revived the ideas in the 19th century.
- In 1897 the electron was discovered, 1909 the nucleus was discovered, 1911 Rutherford suggested that electrons orbit the nucleus, 1919 protons were discovered, and 1932 neutrons were discovered.

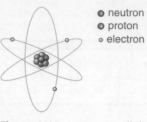

- neutron
- proton
- electron

Figure 1: How can you tell that this atom is not drawn to scale?

Discovering the nucleus

B–A*

- Rutherford's research team fired α-particles at gold foil. Flashes on a fluorescent screen showed where the α-particles hit.
- Most α-particles went straight through, showing that gold atoms are mostly empty space.
- Some were deflected and came out at an angle. To the team's surprise, a few came out backwards.
- Rutherford realised that the α-particles were bouncing off something inside the atoms. He called it the 'nucleus'.

Mass number and isotopes

Properties of isotopes

D–C

- Not all atoms of an element have the same number of neutrons. Atoms of the same element with different masses are called isotopes.
- The proportion of each isotope of an element is fixed, and it is the same in both the pure element and in the compounds containing that element.
- Because isotopes have the same number of electrons, they all react in exactly the same way.

> ### How Science Works
>
> - The relative atomic mass (A_r) of an element is the average mass of an atom. You work it out using the percentage or number of atoms of each different isotope and its mass. Potassium has two isotopes, 90 per cent of the atoms are of potassium-39; the other 10 per cent are of potassium-40. The A_r of potassium is $\frac{(90 \times 39) + (10 \times 40)}{100} = 39.1$

Dating with isotopes

B–A*

- Some elements have radioactive isotopes which decay over time.
- Knowing these and how quickly the isotopes decay, scientists can work out the age of archaeological remains such as Egyptian mummies.
    - Carbon-14 dating is used for remains suspected of being up to 60 000 years old.
    - Older samples need different isotopes. Uranium isotopes were used to calculate the age of the Earth.
    - Fossils are dated by dating the rocks in which they are found.

Improve your grade

Work out the relative atomic mass (A_r) of chlorine. Chlorine has two isotopes, 25 per cent is chlorine-37, 75 per cent is chlorine-35. **AO2 (2 marks)**

Compounds and mixtures

Relative formula mass

- To work out the relative formula mass (M_r) of a compound you find the atomic mass of each element multiplied by the number of atoms of that element present in the formula, then add them up.

- calcium sulfate, $CaSO_4$

  atoms and A_r Ca = 40   S = 32   O = 16

  so   (40 × 1) + (32 × 1) + (16 × 4) = 136

- Atoms and molecules are far too small to be counted. To make counting atoms or molecules easy, chemists call the atomic or formula mass measured in grams a mole.

D–C

More complicated formulae

- A group of atoms sometimes has a set of brackets round it, and a subscript number outside the bracket. You need to take this into account when working out the relative molecular mass.

- For example: aluminium sulfate, $Al_2(SO_4)_3$

  atoms and A_r Al = 27   S = 32   O = 16

  so   (27 × 2) + (32 × 3) + (16 × 12) = 342

Remember!
Atomic masses are either in the question or in the periodic table. In the periodic table, it is the larger of the two numbers in the box for the element!

B–A*

Electronic structure

Electronic structures

- The electrons occupy the lowest available energy levels. They fill up each shell in turn. 2 electrons go in the first (except H which has only 1), up to 8 in the second, the third shell accepts 8 electrons and if the third shell fills up, then the remaining electrons go into the fourth shell – up to calcium.

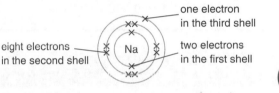

one electron in the third shell

eight electrons in the second shell

two electrons in the first shell

Figure 2: The electronic arrangement of a sodium atom, 2, 8, 1

- For elements beyond calcium the number of outer electrons is the same as the Group number. The number of outer electrons determines the reactivity of an element.

D–C

Noble gas structures

- Noble gases (Group 0) are unreactive because they have a full outer shell of electrons.

- When atoms combine to make compounds, their outer electrons are shared, gained or lost until each atom has a Noble gas electronic configuration.

B–A*

Improve your grade

Use the periodic table to:

a   Work out the electronic structure of these elements: magnesium, chlorine, nitrogen, aluminium, calcium. **AO2 (5 marks)**

b   Work out how many outer electrons each of these elements has: strontium, arsenic, gallium, astatine, germanium. **AO2 (5 marks)**

Ionic bonding

What is ionic bonding?

D–C

- When a metal reacts with a non-metal the metal atoms lose electrons and become positive ions, the non-metal atoms gain electrons and become negative ions.

- A sodium atom (Na) loses an electron to form a sodium ion (Na$^+$) with a 1+ charge. It has become a positive ion. A chlorine atom (Cl) gains an electron to form a chloride ion (Cl$^-$) with a 1– charge. It is now a negative ion.

- The oppositely charged ions are electrostatically attracted. This is ionic bonding.

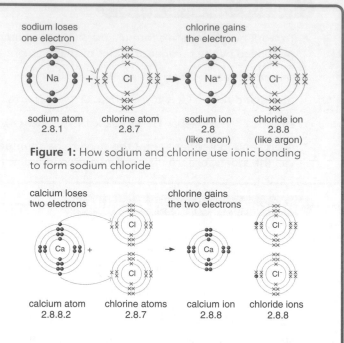

Figure 1: How sodium and chlorine use ionic bonding to form sodium chloride

Figure 2: How calcium forms ionic bonds

Conducting electricity

B–A*

- Ionic compounds can conduct electricity, but only if the ions are free to carry the electric current. Solid sodium chloride does not conduct as the ions cannot move; dissolved in water or molten, it can conduct electricity. This suggests the bonding is ionic.

Alkali metals

Similarities and differences in reactions

D–C

- Group 1 metals react in similar ways because their atoms all have just one electron in their outer shell.
  - lithium just floats on water and fizzes gently
  - potassium bursts into flame, zooms across the water and spits
  - caesium explodes on contact with water.

- The reactivity increases down the Group.

How Science Works

- Trends or patterns, like the reactivity of Group 1 metals, are important. For example, rubidium is below potassium and above caesium in Group 1. So it will probably have a more violent reaction with water than potassium, but less than caesium.

Discovering alkali metals

B–A*

- Group 1 metals, (alkali metals) were discovered in the 19th century. They were too reactive to be separated from their compounds until electrolysis was invented.

Remember!
Reactivity of metals increases as you go down the group, but in non-metals the reactivity increases going up the group.

Improve your grade

Draw diagrams to show how calcium and fluorine lose and gain electrons to make the compound calcium fluoride (CaF$_2$).
AO1 (3 marks)

Halogens

The chemical reactivity of halogens

- This table shows information about Group 7 elements, the halogens.

Table 1: The halogens

name	symbol	description	molecule
fluorine	F	pale yellow poisonous gas	F_2
chlorine	Cl	green poisonous gas	Cl_2
bromine	Br	dark orange–red poisonous liquid that easily vaporises	Br_2
iodine	I	a dark grey solid which on heating becomes a purple gas	I_2

- When halogen elements form compounds fluorine becomes fluoride, bromine becomes bromide, and the ion always has a charge of 1−.

- Fluorine is the most reactive Group 7 element, then chlorine, bromine, and finally iodine which is least reactive.

D–C

Reactive elements, unreactive compounds

- Chlorine is a highly toxic gas. We use it to kill bacteria to make water safe to drink.

- Compounds from chlorine are often very safe and useful:
  - sodium chloride is great on chips, polyvinyl chloride (PVC) is used to make window frames and guttering, not to mention clothes.

Remember!
When a compound is made from two different elements, the compound will have different properties from the original elements.

B–A*

Ionic lattices

Electronic structures

- Solid sodium chloride forms an ionic lattice that is very strong. The oppositely charged ions attract each other in all directions. As a result, each sodium ion is attracted equally to six chloride ions. Each chloride ion is attracted to six sodium ions.

- Melting and boiling points are high, as it is hard to separate the ions from each other. Only when the ions are free to move by melting or dissolving in water can electricity be conducted. Ionic lattices have melting points greater than 500 °C.

- When ionic compounds conduct electricity (electrolysis) they decompose back to their elements. The metal element is formed at the negative electrode or cathode, the non-metal element is produced at the positive electrode or anode.

● Na^+ ◯ Cl^-

Figure 3: The sodium chloride lattice

D–C

Multi-atom ions

- Many negative ions are not single atoms that have lost electrons to become ions, but several non-metal atoms that are covalently bonded together, and have gained electrons as well. For example, the nitrate ion NO_3^-, sulfate ion SO_4^{2-}, and carbonate ion CO_3^{2-}.

- One non-metal ion has a positive charge: this is the ammonium ion, NH_4^+.

Remember!
A compound is likely to have ionic bonding if it has high melting and boiling points, and can only conduct electricity when molten or when in solution.

B–A*

◉ Improve your grade

Explain in detail why sodium chloride:
a   has a high melting point
b   conducts electricity when in solution. **AO1 (3 marks)**

Covalent bonding

Sharing by numbers

D–C

- When atoms of two non-metals combine, they share electrons to achieve noble gas electronic structures, which are stable.

- The outer electron shells of atoms overlap. This sharing forms a covalent bond which holds the two atoms together. Covalent bonds are very strong.

- A chlorine atom has seven outer electrons, so it shares one more from another atom, see Figure 1. It can share from another chlorine atom, or any non-metal atom.

- An oxygen atom has six outer electrons, so needs two more to make eight. In water, the oxygen shares two of its electrons, one with each hydrogen to make H_2O.

- To make an oxygen molecule, oxygen atoms share two electrons with another oxygen atom to make a double bond; two shared pairs of electrons.

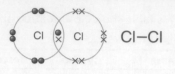

Figure 1: A chlorine molecule

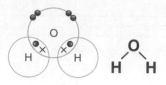

Figure 2: A water molecule

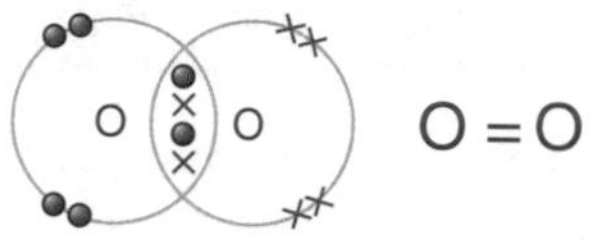

Figure 3: An oxygen molecule

EXAM TIP

When drawing electronic structures to show bonding, to save time and to make it clearer, you can miss out the inner filled electron shells.

How do covalent bonds work?

B–A*

- The nuclei of both atoms involved in a covalent bond are positively charged.

- The electrons involved in the bond are attracted to both nuclei, and so remain a fixed distance apart from each nucleus. This distance is called the bond length.

Covalent molecules

Properties

D–C

- Covalent molecules have no charged particles that are free to move. So they cannot conduct electricity.

- Whilst ionic compounds can dissolve in water, covalent compounds do not.

- Covalent bonds are very strong, but they only exist between the atoms in the molecule. There are weak attractions between the molecules, intermolecular forces. These weak forces are easily broken, so covalent molecules have low melting and boiling points.

Water is odd

B–A*

- Water has a very low relative molecular mass, and should be a gas at room temperature, but water molecules have strong intermolecular forces and so water has high melting and boiling points.

Remember!

Covalent molecules have strong bonds inside the molecule but weak bonds between the molecules. An easy-to-melt, poor conductor of electricity is probably a covalent compound.

Improve your grade

Draw diagrams to show the covalent bonds present in these compounds:
HCl, H_2, CO_2, NH_3, and CH_4. **AO1 (5 marks)**

Covalent lattices

Bonding structure and properties

- Some covalent compounds have giant structures. These structures are called lattices.

- Examples of giant structures include diamond and graphite and silicon dioxide (sand).

- They have very strong structures so the melting and boiling points are very high.

name	structure	hardness	electrical conductor
diamond (carbon)		hardest natural substance	no
graphite (carbon)		soft, used as a lubricant	yes
silicon dioxide (SO_2)		very hard, used as sandpaper	no

D–C

Explaining graphite's properties

- Graphite has layers of atoms that can slide over each other. Three of the four outer carbon electrons form covalent bonds. The fourth is free to move, or delocalised, and can carry an electric current through the structure.

- Each layer is only attracted to the one above and below by weak intermolecular attractions. These allow graphite to be soft and slippery.

Remember!
Pencils leave a mark on paper because the bonds between the layers of graphite are weaker than the bonds between the paper fibres.

B–A*

Polymer chains

Polymers vary

- Polymers are thermosoftening (soften on heating) or thermosetting (harden on heating). Thermosoftening polymers are easy to recycle, thermosetting ones are harder.

- A polymer's use depends on both the monomer and the process used to make it.
  - High density polyethene (HDPE) is used for plastic bottles, and water pipes.
  - Low density polyethene (LDPE) is used to make film and plastic bags.

- HDPE and LDPE are made from the same monomer, but by different catalysts, temperatures and pressures, making different polyethenes with different properties.

- A thermosoftening polymer is a tangle of smooth chains. It is easy for the molecules to slide past each other and melt. A thermosetting polymer has more side chains and links to other chains. It is very hard for the chains to slide at all, so it cannot melt.

D–C

Intermolecular forces

- Polymer molecules are long covalent chains made from carbon atoms, with various side groups attached.

- Intermolecular forces between molecules are weak, but because the molecules are so large the effect is greater, giving them higher melting points than expected.

- Thermosetting polymers soften and melt over a range of temperatures.

B–A*

Improve your grade

Draw diagrams to show the difference between the structure of a thermosetting and a thermosoftening polymer.
AO1 (2 marks)

Metallic properties

Alloys

- Alloys are mixtures of two or more metal elements.

- In a pure metal, all the atoms are the same size. They fit into perfectly regular layers which can slide easily. Larger or smaller atoms in alloys disrupt the regular lattice. The layers slide less easily, so alloys are harder and less malleable than pure metals.

- Most metals in common use are alloys, because they are stronger than pure metals and have lower melting points.

- Special alloys have been developed that can remember their shape.

- These shape memory alloys such as nitinol are useful, as when warmed they return to their original shape. Uses include dental braces; and plates to hold broken bones together.

- In metals, the outer electrons of each atom are delocalised or free to move.

- As the sea of electrons can move, the metal is able to conduct electricity.

- The structure allows metals to bend, as the positive ions can move around each other.

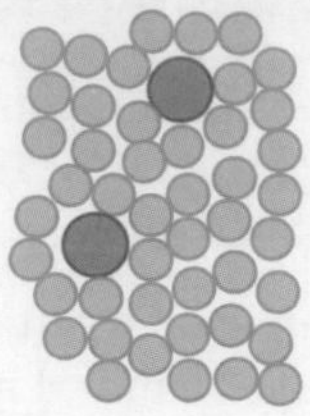

D–C

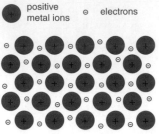

Figure 1: An alloy lattice

positive metal ions   electrons

Figure 2: Metallic bonding

Conduction

B–A*

- When a potential difference is applied to a metal, the delocalised electrons flow from the negative connection towards the positive connection creating an electrical current.

- Heating a metal makes the particles in it vibrate more. The vibrations pass along the metal lattice, 'warming' the metal so metals are good conductors of heat.

Modern materials

Applications of modern materials

D–C

- Modern materials include 'Smart' materials, such as photochromic, thermochromic, and shape memory alloys. There are also very small materials known as nanoparticles.

- Photochromic materials change colour according to the intensity of light.

- Thermochromic materials change colour according to the temperature.

- Nanoparticles are very small, being from 1–100 nm in size, with up to 300 atoms.

- Carbon has several nanoparticles, such as nanotubes; and Buckminsterfullerene (C_{60}).

- Uses of smart materials include self-cleaning and shading glass.

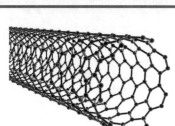

Figure 3: A carbon nanotube

Fullerenes

B–A*

- Buckminsterfullerene (C_{60}) is a fullerene. They are based around five or six carbon-atom rings joined together in a geodesic sphere. It is possible to trap a molecule inside the sphere, so allowing highly toxic drugs to be carried to specific sites in the body to kill infections or cancerous cells.

Remember!

Nanoparticles have an enormous surface-area-to-volume ratio, so they make excellent catalysts.

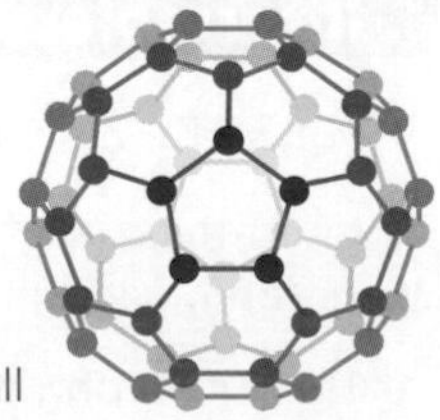

Figure 4: A buckminsterfullerene ball

Improve your grade

Explain how the metal lattice allows for metals to be stretched without breaking. **AO1 (2 marks)**

Identifying food additives

Chromatography

- Paper chromatography shows that the colourings in a food are those listed on the label.

- When a solvent rises up the paper, it carries the colours with it and separates them.

- Each chemical in a mixture travels up the paper at the same speed as it does when alone.

- The chromatogram in Figure 5 shows that the food contains E133 and E102, as the food has spots at the same height as the samples. It does not have any E131 and E142, but it has a third chemical that is not identified.

- Chromatography is a separating technique also used with drugs and medicines.

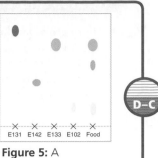

Figure 5: A chromatogram of food colours

D–C

Further chromatography

- To save time in analysing chromatograms, chemists use the retention factor (R$_f$).

- Any chemical will always travel the same distance up a standard paper in the same time in the same solvent. The retention factor can be calculated like this:

$$R_f = \frac{\text{distance moved by the substance}}{\text{distance moved by the solvent front}}$$

- Thin-layer chromatography uses a glass plate coated with a thin layer of absorbent materials. It is quicker than paper chromatography, but also more expensive.

B–A*

Instrumental methods

Retention times and mass spectrometry

- Analytical chemists use a technique called gas chromatography. They send a gas solvent with the substances to be analysed, through a tube packed with a solid material.

- The components travel through the tube at different speeds and are detected at the end. The results are shown as a graph, with each peak representing a component and the height of the peak indicating how much was present.

- The position of the peak is used to work out the retention time for each component.

- Often a mass spectrometer is attached to the gas chromatography equipment, a GC-MS method. This measures the relative atomic or molecular mass of each component.

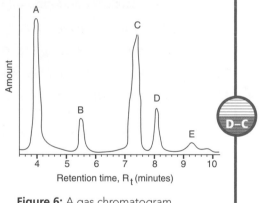

Figure 6: A gas chromatogram retention–time graph

D–C

How does mass spectrometry identify substances?

- The mass spectrometer knocks an electron of each atom or molecule, leaving a positively charged particle or molecular ion. It can then measure the mass of the particle.

- It also breaks up some of the molecules into smaller pieces, and this helps to distinguish between different molecules with the same relative molecular mass.

B–A*

Improve your grade

Look at Figure 6. It shows the retention–time graph for a gas chromatogram sample.
a   How many different components were in the sample?
b   List the components in order of the quantity of each present. Start with the most plentiful. **AO3 (2 marks)**

Making chemicals

Routes and conditions

D–C

- There are several types of chemical reaction that can be used to make a chemical:
  - oxidation or reduction, losing or gaining oxygen
  - neutralisation, reacting an acid with an alkali or a base
  - precipitation, reacting two soluble compounds to make an insoluble product
  - electrolysis, using electricity to split a liquid or solution.
- Whichever type of reaction is used, the reaction conditions are important in terms of speed, costs, and safety:
  - temperature, higher temperatures make reactions go quicker
  - concentration, more concentrated solutions react faster
  - pressure, higher pressure compresses gas molecules, giving more molecules in the same volume and leading to faster reactions
  - particle size, smaller pieces react faster than larger pieces
  - catalysts, these can speed up the reaction rate, and are unchanged by the reaction.

Chemical calculations

B–A*

- Chemists calculate the correct masses of reactants to use to make products, so that only the smallest amounts needed are used to make the products and waste products are kept to the minimum.
- This keeps costs to the minimum and reduces waste.

Chemical composition

Empirical formulae

The relative quantities of each element present in a compound can be worked out using the percentage composition method:

- find the relative formula mass (M_r), for example $FeSO_4$ = 152
- find the mass of the element in the compound, Fe = 56 × 1, = 56
- divide 56 by M_r and multiply by 100 to get the percentage of iron in iron sulfate = 39.4%.

Working out a formula from experimental data

D–C

To calculate the formula of copper oxide from an experiment, you need to know: how much copper was used, how much copper oxide was produced, how much oxygen was needed.

If 1.27 g of copper was burnt, and made 1.59 g of copper oxide, what is the formula of copper oxide?

To answer this, you need to:

- calculate the mass of oxygen used as 1.59 g – 1.27g = 0.32 g
- divide the mass of copper by its A_r (1.27/63.5 = 0.02) and oxygen by its A_r (0.32/16 = 0.02)
- divide each result by the smallest one, i.e. 0.02, to give the ratio of atoms in the formula, so Cu = 0.02/0.02 = 1, O = 0.02/0.02 = 1, so the formula is CuO.

Analysing hydrocarbons

B–A*

You can analyse hydrocarbons using the same method. Burning 1.68 g of hydrocarbon produced 5.28 g of carbon dioxide, and 2.16 g of water. So:

- mass of carbon present (12/44 × 5.28 = 1.44 g), mass of hydrogen present (2/18 × 2.16 = 0.24 g)
- divide each mass by its A_r C = 1.44/12 = 0.12, H = 0.24/1 = 0.24
- ratio of atoms is 0.12 : 0.24, or 1:2, so the empirical formula is CH_2, so it is an alkene.

| A_r, H = 1, C = 12, O =16 |
| M_r, CO_2 = 44, H_2O = 18 |

Improve your grade

Calculate the formula of lithium oxide made from 5.6 g of lithium and 6.4 g of oxygen. **AO2 (2 marks)**

Quantities

Quantities from equations

- Look at this equation for the production of ammonia from nitrogen and hydrogen:
  $N_2(g) + 3H_2(g) \rightleftharpoons 2NH_3(g)$. The equation shows that one nitrogen molecule reacts with three hydrogen molecules to produce two molecules of ammonia gas.

- We can then work out the **relative reaction masses** of each reactant or product by multiplying the A_r or M_r by the number of reacting molecules for each gas.
  $N_2 = 28 \times 1 = 28$, $H_2 = 2 \times 3 = 6$, and $NH_3 = 17 \times 2 = 34$.

- 28 units of nitrogen will react with 6 units of hydrogen to make 34 units of ammonia. To work out how much ammonia can be made if 56 tonnes of nitrogen is used, you divide 56 by 28 to find the reacting amount (= 2); so 56 tonnes of nitrogen will make 34×2 tonnes of ammonia = 68.

EXAM TIP

Whenever there is an important or difficult calculation required that needs a balanced equation, the balanced equation will be given in the question.

D–C

Sir Terry Pratchett's sword

- If a good copper ore contains 5% by mass of copper carbonate. How much copper could be made from 25 kg of ore?

- $2CuCO_3(s) + C(s) \rightarrow 2Cu(s) + 3CO_2(g)$

- the M_r for $CuCO_3$ is $63.5 + 12 + (16 \times 3) = 123.5$
  - reacting mass for $CuCO_3$ is $123.5 \times 2 = 247$
  - reacting mass for copper is $63.5 \times 2 = 127$
    Assuming the 25 kg of ore is used and contains $(5/100) \times 25 = 1.25$ kg of copper carbonate, the possible yield will be $(1.25 \times 127)/247 = 0.64$ kg

A_r	
Cu = 63.5	
O = 16	
C = 12	

B–A*

How much product?

Calculating percentage yield

- Other factors affect the amount of product produced in a chemical reaction.

- Reactants are rarely 100% pure. The mass of actual reactant is less than the mass weighed out, so forms less product: the same reactants can form different products: if the reaction is reversible, some of the products turn back into reactants.

- To compare the effectiveness of making a chemical in different ways, we calculate the percentage yield

$$\text{percentage yield} = \frac{\text{actual yield}}{\text{theoretical yield}} \times 100$$

Remember!

When calculating percentage yields, it doesn't matter what units are used for the actual yield or the theoretical yield so long as they are both measured in the same units.

D–C

Percentage yields, economics and the environment

- Percentage yield indicates how effectively reactants are converted into the product.

- It is usually more economic to use a reaction with a high percentage yield.

- Maximising percentage yields makes economic sense. Less raw materials are needed and less are wasted in making unwanted by-products that have to be disposed of.

B–A*

Improve your grade

Calculate the mass of calcium oxide that can be made from 100 tonnes of calcium carbonate.
The balanced equation for the reaction is:
$CaCO_3(s) \rightarrow CaO(s) + CO_2(g)$ **AO2 (3 marks)**

Reactions that go both ways

Examples of reversible reactions

D–C

- Heating solid ammonium chloride causes it to decompose into two gases, hydrogen chloride and ammonia. When the two gases cool, the ammonium chloride is reformed. This is a reversible reaction.

- Copper sulfate crystals are blue, but heat them and they turn white, as the water of crystallisation is removed. Add a little water to the white anhydrous copper sulfate and the white powder turns blue again.

$$CuSO_4.5H_2O(s) \rightleftharpoons CuSO_4(s) + 5H_2O(\ell)$$

- The double-headed arrow shows that the reaction can go both ways, left to right or forward, right to left or backward.

Reversible reactions used in industry

B–A*

- Ammonia is made using a reversible reaction:

$$N_2(g) + 3H_2(g) \rightleftharpoons 2NH_3(g)$$

- Sulfuric acid is needed to make many other chemicals. It too is manufactured using two reversible reactions.

$$2SO_2(g) + O_2(g) \rightleftharpoons 2SO_3(g) \quad \text{then,}$$

$$SO_3(g) + H_2O(\ell) \rightleftharpoons H_2SO_4(\ell)$$

How Science Works

- When investigating reversible reactions and the effect of different conditions upon them, it is really important to be clear what the independent variable and control variables are in the investigation.

Rates of reaction

Measuring rates of reaction

D–C

- When chemists talk about the rate of reaction they mean how much chemical reacts, or is formed, in a given time.

- rate of reaction = $\dfrac{\text{amount of reactant used up}}{\text{time taken}}$ or $\dfrac{\text{amount of product formed}}{\text{time taken}}$

- You can alter the rate of a chemical reaction by doing one or more of these: increasing the temperature, increasing the concentration of a reactant, using smaller pieces of a reactant, or adding a catalyst.

- Often a gas is produced and so it is easy to measure the quantity produced. Figure 1 shows ways to measure a gas.

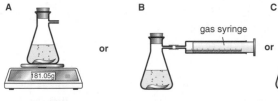

Figure 1: Ways to measure the quantity of a gas produced

Calculating reaction rates

B–A*

- You can either calculate the average reaction rate for a reaction; or the reaction rate at a given time.
  - To calculate the average reaction rate from the graph, you can see that in 4 minutes the reaction produced 12.8 cm³ of hydrogen gas. So 12.8/4 = 3.2 cm³ per minute.
  - To calculate the rate at 2–2.5 minutes, use the graph to find the change in volume from 2 to 2.5 minutes. This is 0.8 cm³, the time for this change to happen is 0.5 minutes, so 0.8/0.5 = 1.6 cm³ per minute.

Figure 2: Measurements from a magnesium–acid reaction

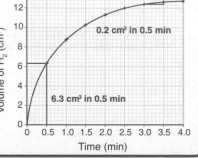

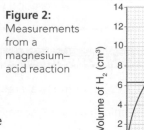

Improve your grade

Use Figure 2 to calculate the rate of the reaction at the following times:
a   0 to 0.5 minutes
b   1.0 to 2.0 minutes
c   2.5 to 3.0 minutes. **AO3 (3 marks)**

Collision theory

Explaining collision theory

- For a chemical reaction to occur, the reactants must collide with each other. Any change that increases the number of collisions will increase the rate of a reaction. Not all collisions cause a reaction. The collision has to be hard enough to cause the reactants to react. This can be done by increasing the temperature, increasing the concentration, increasing the pressure, using smaller particles of a solid, adding a catalyst.

● reactant particles ✳ successful collision

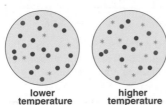

| lower temperature | higher temperature |

Figure 3: Effect of temperature on the number of successful collisions per second in a liquid or gas

| lower concentration | higher concentration |

Figure 4: Effect of concentration on the number of successful collisions per second in a liquid or gas

| large particle | smaller particles |

Figure 5: Effect of surface area of a solid on the number of successful collisions per second

D–C

How much change?

- Doubling the concentration of one reactant often doubles the rate.

- Cutting a cube of solid reactant into eight, doubles its surface area and doubles the rate.

- Increasing the temperature always increases the rate.

- Different reactions need different catalysts. Not all reactions can be catalysed.

Remember!
Not all collisions cause reactions.

B–A*

Adding energy

Investigating the effects of temperature

- Magnesium reacts with sulfuric acid, giving off hydrogen gas.

$$Mg(s) + H_2SO_4(aq) \rightarrow MgSO_4(aq) + H_2(g)$$

- Figure 6 shows results from using the same amounts of magnesium and acid at three different temperatures.

- The steeper the graph's curve, the faster the rate. All three reactions produce the same final volume of hydrogen gas, as they have the same quantities of starting reactants.

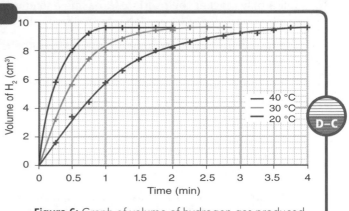

— 40 °C
— 30 °C
— 20 °C

Figure 6: Graph of volume of hydrogen gas produced at different temperatures

D–C

Activation energy

- To cause a reaction, particles must collide with sufficient energy to break bonds.

- The minimum energy needed for this to happen is called the activation energy.

B–A*

Improve your grade

A student was investigating a chemical reaction. She decided to increase the temperature of the reaction, whilst decreasing the concentration of one reactant. Use collision theory to explain why the rate of reaction remained unchanged. **AO3 (2 marks)**

Concentration

The effect of concentration

- Higher concentration means that there are more reactant particles in the same volume. There are more collisions per second, increasing the reaction rate. Increasing the concentration does not affect the energy of collisions.

- Reactants are used up during a reaction and their concentrations decrease, so the reaction slows down. That is why the graphs are curves.

- To find the rate at any point during the reaction find the gradient of the graph at that point.

- Look at Figure 1. What is the rate at 0.5 mol/dm³ after 2 minutes?
  - Draw a tangent at this point, and construct a triangle as shown.
  - Find the scale lengths on each axis: x-axis = 2 min, y-axis = 3.3 cm³
  - Rate = amount of product ÷ time = 3.3/2 = 1.65 cm³ of H₂/min

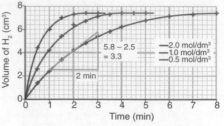

Figure 1: Rate-of-reaction graphs for magnesium reacting with sulfuric acid at different concentrations

Two solutions

- Doubling the concentration of one reactant will double the rate of the reaction. If you double both reactants then the reaction rate will double twice, that is go four times as fast and be complete in a quarter of the time.

How Science Works

- A chemical reaction is complete when there are no more reactants available to react. It may stop because all the reactants have been used up or, more likely, when just one of the reactants has been used up. The remaining reactant is said to be in **excess**.

Size matters

Size and surface area

Solids only react at their surface. Breaking a solid into smaller pieces exposes more surface area. That provides more reactant particles to be available for collisions: the reaction rate increases.

The ratio between the two surface areas shows the change in rate of reaction. For example, if you reduce particle size from 1 cm³ to 1 mm³:

- for 1 cm cube surface area:
  - each face = 10 mm × 10 mm = 100 mm²
  - total = 6 × 100 mm² = 600 mm²

- for 1 mm cube surface area:
  - each face = 1 mm × 1 mm = 1 mm²
  - total = 6 × 1 mm² = 6 mm²

- A 1 cm cube contains 1000 cubes of 1 mm side length, so:
  - surface area of 1000 cubes = 6000 mm²
  - 6000/600 = 10, so the reaction will be ten times faster.

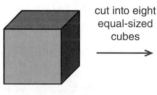 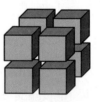

cut into eight equal-sized cubes

Figure 2: Smaller particles, larger surface area

Liquids and gases

- Liquids react with gases much faster if sprayed into tiny droplets first.

- Oil-fired boilers and furnaces spray in the fuel oil, so that it burns rapidly.

Improve your grade

Use Figure 1 to find the rate of the reaction after 1 minute when the concentration is 2 mol/dm³. **AO2 (2 marks)**

Clever catalysis

Using catalysts

- Catalysts are chemicals that speed up reactions, but are not used up in the process. They can be used over and over again.

- Car exhaust fumes contain poisonous carbon monoxide and oxides of nitrogen. In a catalytic converter platinum, rhodium and palladium, catalyse reactions between the gases producing less harmful ones, for example carbon monoxide becomes carbon dioxide, nitrogen oxides are converted to nitrogen. Many catalysts are transition metals or their compounds.

Table 1: Some catalysts and their uses

catalyst	process	product
iron	Haber Process	ammonia
platinum	oxidising ammonia	nitric acid
nickel	hydrogenating vegetable oils	vegetable spreads such as margarine

D–C

How catalysts work

- A catalyst provides a way for a reaction to occur with lower activation energy. So more collisions now result in reactions, and the rate is increased.

- The reactants do not react directly. One combines with the catalyst. This combination reacts with the second reactant, making the product and regenerating the catalyst.

- When another chemical reacts with the catalyst it can be poisoned, and stop working.

- Enzymes are biological catalysts. Like other catalysts, they are specific to reactions and we each have hundreds of different enzymes working all the time in our bodies to keep us alive. They are also used in biological washing powders, brewing and baking.

Figure 3: Each circle represents a reactant. The catalyst reacts and changes, but is regenerated again

B–A*

Controlling important reactions

Economics and safety

- When planning an industrial process, chemists need to consider all the variables that affect rates of reaction. Making ammonia is a good example:

- $N_2(g) + 3H_2(g) \rightleftharpoons 2NH_3(g)$
    - the temperature used is 450 °C
    - the pressure used is 200 atmospheres (this has the effect of increasing the concentration)
    - an iron catalyst is used
    - the catalyst has a large surface area.

- The choice of conditions produces a small yield of ammonia (15–20%), but quickly and cheaply. The unreacted nitrogen and hydrogen are fed back in, so eventually all of the reactants become ammonia.

- Increasing temperature speeds up the reaction, but reduces the amount of ammonia formed, the high pressure produces more ammonia, but is very expensive, the iron catalyst speeds up the reaction, whilst being cheap.

- Many chemical reactions are exothermic, they produce heat. This can alter the rate of the reaction so the chemical plant needs to be able to keep the temperature constant.

D–C

Catalysts in nature

- Animals and plants use catalysts. Photosynthesis relies upon chlorophyll as a catalyst to convert carbon dioxide and water quickly to sugars.

- For plants and animals to respire, special biological catalysts – enzymes – are needed to change the sugar back to carbon dioxide, water and the energy we need.

B–A*

Improve your grade

What is activation energy? Explain how a catalyst lowers the activation energy for a particular reaction. **AO1 (3 marks)**

The ins and outs of energy

Energy stores

- All substances store energy, but different substances store different amounts.

- If reaction products store less energy than the reactants did, the reaction is exothermic. The extra energy is transferred to the surroundings, heating them up.

- If the products store more energy (an endothermic reaction), they must have absorbed energy from somewhere. The surroundings provide the energy by cooling down.

- These changes can be shown by an energy-level diagram.

- Exothermic and endothermic reactions are useful. Calcium oxide reacts exothermically with water to provide a heat source for self-heating cans. Ammonium nitrate is used to provide an endothermic reaction to make sports injury packs, to cool muscles.

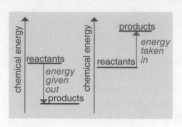

Figure 1: Exothermic and endothermic energy-level diagram

Figure 2: Self-heating coffee can

D–C

Energy and reaction rates

B–A*

- Energy changes affect temperature, which affects reaction rates.

- Endothermic reactions cool, so to maintain the rate they need to be supplied with heat.

- Exothermic reactions produce heat, so need to be cooled down if the rate is not going to increase.

- With reversible reactions, if the forward reaction is endothermic, then the backward reaction will be exothermic.

Acid-base chemistry

Acid-base reactions

D–C

- Most bases are insoluble in water. Bases that do dissolve in water are alkalis, because their solutions contain hydroxide ions in the solution. Alkalis are usually hydroxides of Group 1 and 2 metals in the periodic table.

- A base reacts with an acid to form a salt and water, for example zinc oxide and sulfuric acid.

$$ZnO(s) + H_2SO_4(aq) \rightarrow ZnSO_4(aq) + H_2O(\ell)$$

- Metal carbonates also produce carbon dioxide.

$$CaCO_3(s) + 2HCl(aq) \rightarrow CaCl_2(aq) + H_2O(\ell) + CO_2(g)$$

- Alkalis react with acids to give a salt and water only.

- Group 1 carbonates and hydrogencarbonates are bases because they react with acids to give salt and water, and carbon dioxide.

- Neutralisation reactions are exothermic and can be represented by this ionic equation:

$$H^+(aq) + OH^-(aq) \rightarrow H_2O(\ell)$$

Ammonia

B–A*

- Ammonia solution is alkaline because when it dissolves in water it becomes ammonium hydroxide solution.

- If ammonia reacts directly with an acid, such as hydrochloric acid, it does not produce water. Instead, it produces the salt.

$$NH_3(g) + HCl(aq) \rightarrow NH_4Cl(aq)$$

Improve your grade

Sketch an energy-level diagram for the exothermic reaction that takes place when nitric acid reacts with sodium hydroxide. Write on the line for 'reactants', the names of the reactants; and on the line for 'products', the products. **AO1 (4 marks)**

Making soluble salts

Practical methods

- Soluble salts can be made from acids in one of three ways: reacting with a metal, reacting with an insoluble base, reacting with an alkali. The name of the salt that's made depends on the metal, or on the metal in the name of the alkali or base, and the acid.

- To make a salt from a solid: add the solid to some dilute acid until no more will dissolve, if the reaction is slow, then warm the mixture gently, filter the solution to remove unreacted solid, evaporate the filtered solution until it is nearly all gone, then allow it to cool and the salt crystals will appear.

- To make a salt using an alkali: place 25 cm³ of alkali in a conical flask, and add some universal indicator, add the acid from a burette, syringe or measuring cylinder until the acid is neutralised – note the volume of acid required, then either repeat the experiment omitting the universal indicator; or add a little powdered charcoal to the coloured solution and heat, then filter to remove the charcoal, evaporate the solution until it is nearly all gone.

Table 1: Some acids and their salts

acid		salt made
sulfuric	H_2SO_4	sulfate
hydrochloric	HCl	chloride
nitric	HNO_3	nitrate

Remember!

The higher a metal is in the reactivity series, the easier it will react with the acid. At the top of the reactivity series the reaction will be extremely violent – so Group 1 salts are made by other methods.

Neutralisation may not work

- Not all salts are soluble. Some metals either have a surface layer of oxide that prevents them from reacting; or the reaction with the acid coats them with the salt, preventing the rest of the metal atoms reacting with the acid.

EXAM TIP

If asked to make a salt, first decide on the acid you need, then on a suitable source of the metal.

Insoluble salts

Useful precipitations

- An insoluble salt is one that will not dissolve in water. Insoluble salts are made by mixing together two solutions, each containing one part. The metals swap partners. The solid product formed by mixing two solutions is called a precipitate. Filtering removes the precipitate, which is washed before drying in an oven.

- Salts are ionic compounds. A mixture of two salts in solution contains positive ions of two metals and negative ions of two non-metal groups. If any pair combines to form an insoluble salt, they produce a solid precipitate.

- Precipitation reactions are useful to remove unwanted ions from water; calcium ions cause hardness in water and are removed using sodium carbonate solution, phosphate ions in washing powder are removed by adding aluminium ions to the waste water. Calcium carbonate is produced for toothpaste this way.

Chemical analysis

- Sulfates can be identified by adding barium chloride solution. A white precipitate of barium sulfate appears.

- Halogen compounds can be identified by adding silver nitrate.

- Sodium hydroxide can identify metal ions in solution.

Improve your grade

Describe the method to make some solid chromium chloride from chromium metal. **AO3 (4 marks)**

Ionic liquids

Conduction in liquids

- Pure water does not conduct electricity as it is made of molecules. Tap water is not pure as it contains many dissolved ions, so it does conduct electricity.

- When a liquid conducts, the electric current must enter and leave the liquid through solid conductors called electrodes. This is electrolysis.

- The ions in a solution or molten compound move around randomly. In electrolysis, positive ions are attracted to the negative cathode and negative ions are attracted to the positive anode.

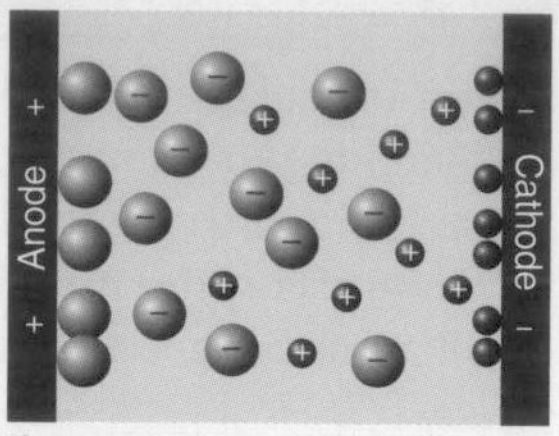

- As the ions move, they carry electric charge through the liquid from one electrode to the other.
  - At the anode, the negative ions release their electrons; whilst at the cathode, the positive ions pick up electrons at the same rate as the negative ions are releasing their electrons.
  - Metal ions gain electrons at the cathode, and become atoms that coat the electrode or react with the liquid. Non-metal ions at the anode lose electrons, and may be released as a gas or react with the liquid.

- Electroplating is another use of electrolysis (see Figure 2). An object is coated with a thin layer of metal, as shown. The silver anode dissolves into the solution, maintaining the concentration of silver ions.

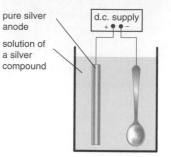

Figure 1: Ions in a solution that is conducting electricity

Figure 2: Electroplating a spoon

Direct and alternating current

- Electrolysis only takes place with direct current (d.c.), mains electricity is alternating current (a.c.), so does not allow electrolysis; but the water still conducts electricity.

Remember!

As electrolysis of a solution continues, the concentration of the compound in the solution (electrolyte) is reduced, so more of the compound needs to be added. In electroplating, the anode is often made from the metal to be plated; it dissolves into the electrolyte so maintaining the metal ion's concentration.

Electrolysis

More applications of electrolysis

- Electrolysis of molten sodium chloride decomposes the compound producing sodium metal at the cathode and chlorine gas at the anode.
  - Electrolysis of a solution of sodium chloride (brine) is slightly different. The water makes the brine a mixture of ions (Na^+, H^+, OH^- and Cl^-). Chlorine gas is produced at the anode, at the cathode, the lower of sodium and hydrogen (the positive ions) in the reactivity series is made, which is hydrogen; and the sodium ions react with the hydroxide ions to make a solution of sodium hydroxide.

- Aluminium metal is obtained by electrolysis from aluminium oxide (Al_2O_3) dissolved in cryolite to reduce the melting point of the oxide.

- Copper is purified by electrolysis.

Half-equations

- Two reactions happen during electrolysis, one at each electrode. These can be represented as **half-equations**.

- In sodium chloride electrolysis:
  - at the cathode, $Na^+ + e^- \rightarrow Na$ (this is reduction)
  - at the anode, $2Cl^- \rightarrow Cl_2 + 2e^-$ (this is oxidation)

Improve your grade

Write half-equations to show the reactions, at both the anode and cathode, for the electrolysis of copper chloride ($CuCl_2$). **AO2 (2 marks)**

Structure and bonding

Chemical bonding involves either transferring or sharing electrons in the outer shells of atoms.

Ionic compounds are held together by strong electrostatic forces of attraction.

Atoms that lose electrons become positively charged ions. Atoms that gain electrons become negatively charged ions.

When atoms share pairs of electrons, they form covalent bonds.

Structure, properties and uses

When melted or dissolved in water, ionic compounds conduct electricity. Covalent compounds do not.

Ionic compounds have regular structures (giant ionic lattices), high melting and boiling points.

Diamond and graphite have different properties determined by their structures.

Metals consist of giant structures of atoms arranged in a regular pattern.

The properties of thermosoftening and thermosetting polymers depend on what they are made from and how they are made.

Nanoscience refers to structures that are 1–100 nm in size and of the order of a few hundred atoms.

Atomic structure, analysis and quantitative chemistry

Atoms of the same element can have different numbers of neutrons.

Elements and compounds can be identified using instrumental methods.

The relative atomic mass (A) and the relative formula mass (M) allow numbers of particles to be compared.

The masses of reactants and products can be calculated from balanced symbol equations.

The amount of a product obtained is known as the yield.

A reversible reaction is one where the products of the reaction can react to produce the original reactants.

Rates of reaction

Catalysts change the rate of chemical reactions but are not used up during the reaction.

The minimum amount of energy that particles must have to react is called the activation energy.

The rate of reaction $= \dfrac{\text{amount of reactant used}}{\text{time}}$ or $= \dfrac{\text{amount of product formed}}{\text{time}}$

Chemical reactions can only occur when reacting particles collide. Collision theory explains why changes to conditions affect rates.

Exothermic and endothermic reactions

An exothermic reaction transfers energy to the surroundings. An endothermic reaction takes in energy from the surroundings.

If a reversible reaction is exothermic in one direction, it is endothermic in the opposite direction.

Acids, bases and salts

Metal oxides and hydroxides are bases. Soluble hydroxides are called alkalis. Ammonia makes an alkaline solution to produce ammonium salts.

Soluble salts can be made from acids by reacting them with metals, insoluble bases, or alkalis. Salt solutions can be crystallised to produce solid salt. Insoluble salts can be made by mixing solutions of ions so that a precipitate is formed.

In neutralisation reactions, hydrogen ions react with hydroxide ions to produce water.

Hydrogen ions, $H^+(aq)$, make solutions acidic and hydroxide ions, $OH^-(aq)$, make solutions alkaline.

Electrolysis

When an ionic substance is melted or dissolved in water, the ions are free to move about within the liquid or solution.

During electrolysis, positively charged ions move to the negative electrode and are reduced, and negatively charged ions move to the positive electrode and are oxidised.

Aluminium is manufactured by the electrolysis of a molten mixture of aluminium oxide and cryolite.

Reactions at electrodes can be represented by half-equations, for example:
$2Cl^- \rightarrow Cl_2 + 2e^-$ or $2Na^+ + 2e^- \rightarrow 2Na$

The electrolysis of sodium chloride solution produces hydrogen, sodium hydroxide and chlorine.

Ordering elements

Breaking the code

- John Newlands and Dmitri Mendeleev both used the idea of 'octaves' to help them develop the periodic table.

- They both ordered the elements by mass – but Mendeleev left gaps for undiscovered elements, which enabled him to maintain the pattern of properties, and predicted the properties of some of the undiscovered elements. He also realised that iodine should come after tellurium, based on chemical properties, despite being lighter in mass.

- When eka-aluminium (gallium) was discovered in 1875, and matched Mendeleev's predictions about it, the periodic table was accepted.

D–C

EXAM TIP

Mendeleev did not know about atomic numbers, as protons had not been discovered. Fortunately for Mendeleev, none of the noble gases had been discovered when he suggested the periodic table. If some of them had been, it would have made it harder to come up with the table.

Noble gases

- The noble gases were discovered between 1868 and 1898. As they were all in the same group, their chemistry was similar, so they were added as Group 0 at the right-hand side of the periodic table.

B–A*

Remember!
Mendeleev succeeded because he realised that science did not know everything, and made allowances for this.

The modern periodic table

Electron configurations

- Elements in the same group have the same number of outer electrons, this is the group number. This is why their chemistry is similar.

D–C

Figure 1: Electron configurations. Note how filling a new energy level starts a new period in the periodic table. What other patterns can you see?

Work in progress

- Elements 1 to 92 occur naturally on Earth, either as the pure element or in a compound.

- Elements from 93 onwards are made inside nuclear reactors, more elements are being added slowly, with element 118 made in 2006. The three atoms of this lasted less than a second before decaying radioactively.

B–A*

⦿ Improve your grade

Describe the two assumptions Mendeleev made when designing his periodic table. How did these assumptions lead to his table being accepted? **AO1 (4 marks)**

Group 1

Explaining the trends

- You already know that the Group 1 metals react easily with water, and that the reaction gets more violent the further down the group you go.

- All Group 1 metals have one electron to lose. As the element gets lower in the group, the distance from the nucleus to this outer electron gets larger and so it is easier for the negatively charged electron to escape the positive nucleus. Also, there are more shells of electrons between the outer electron and the nucleus, which reduces the positive nucleus's effect; so the further down the group, the easier the electron is to lose so the more violent the reaction.

D–C

Francium

- Francium is the most reactive of the Group 1 metals. It is highly radioactive and has no commercial uses.

How Science Works

- Mendeleev used trends or patterns in chemical properties to develop the periodic table. Trends in Group 1 help us to develop theories to explain why the reactions are increasingly violent.

B–A*

Transition metals

Electronic configuration and properties

- Transition metals are typical metals, but they also form a wide range of coloured compounds used in making paints. The colours are caused by their electronic structures.

- From titanium onwards in the periodic table, more electrons are added to the third shell, giving copper an electronic structure of 2.8.17.2.

- Transition metals can alter their electron arrangements to form several different ions, for example Fe^{2+} and Fe^{3+}. Another example, vanadium, has four different possible ions.

- Transition metals are useful as catalysts in many reactions, such as making ammonia (NH_3).

D–C

Highly desirable and rare

- Many common metals are transition metals, for example, iron, steel and copper.

- There are also some transition metals that are useful but very rare: silver, gold and platinum are three of them.

B–A*

Improve your grade

Explain why potassium is a more reactive metal than lithium. **AO1 (2 marks)**

Group 7

Explaining the trends

D–C

- Just like Group 1 metals, Group 7 – the Halogens – change their reactivity as you go down the group. Unlike Group 1, they get less reactive further down the group.

- The reactivity is so great that if a solution of chlorine is added to a solution containing either bromide or iodide ions, the chlorine will displace the bromide or iodide, becoming chloride ions and leaving the bromine or iodine in the solution. For example:

$$2NaI(aq) + Cl_2(aq) \rightarrow 2NaCl(aq) + I_2(aq)$$

- The smaller the halogen atom, the more strongly the positive nucleus is felt because there are few shells of electrons to shield the positive charge. Electrons are easily attracted. Larger halogen atoms have a greater distance to the nucleus, so the positive attraction is less, and it is also shielded by more electron shells. Small (top of the group) halogen atoms are more reactive than larger (lower in the group) atoms.

> **EXAM TIP**
>
> When explaining the increased reactivity of Group 1 metals down the group, or the decreased reactivity down the halogens, the same two factors apply: the distance from the outer electron shell to the nucleus, and the effect of the number of shells in the atom. The only difference is whether the atom wants to lose an electron (Group 1) or gain an electron (Group 7).

Same formula, different bonding

B–A*

- Hydrogen chloride is a covalently bonded molecule when a gas. Dissolve it in water and it becomes hydrogen and chloride ions. This makes the solution hydrochloric acid owing to the hydrogen ions: $H^+(aq)$.

> **Remember!**
>
> With displacement reactions of halogens, the element higher up the group is more reactive and takes the extra electron from the halide ion to become an ion itself; leaving the lower element as molecules.

Hard and soft water

Water softening

D–C

- Hardness in water is caused by the presence of calcium or magnesium ions dissolved in the water. Permanent hard water has dissolved calcium or magnesium sulfate, nitrate or chloride. Temporary hard water has dissolved calcium or magnesium hydrogencarbonate.

- Temporary hardness can be removed by boiling the water, changing the hydrogencarbonate ions to carbonate ions, and releasing carbon dioxide gas. Permanent hard water needs either the addition of sodium carbonate (washing soda) or to be passed through an ion exchange column that replaces the calcium or magnesium ions with sodium or hydrogen ions.

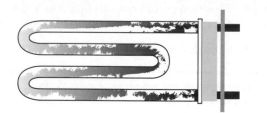

Figure 1: Limescale makes boilers, kettles and washing machines inefficient, so they use more electricity. Eventually, the build-up of limescale causes appliances to break down

Ionic equations

B–A*

- Reacting hard water with soap forms a scum. This is a calcium salt of the soap, and is insoluble.

- Adding sodium carbonate precipitates out calcium or magnesium carbonate:

$$Na_2CO_3(aq) + MgSO_4(aq) \rightarrow MgCO_3(s) + Na_2SO_4(aq)$$

Improve your grade

Explain why permanent hard water cannot be softened by boiling yet temporary hard water can. **AO2 (3 marks)**

Safe drinking water

Water filters

- Tap water is not pure water as it contains dissolved solids. Water filters attempt to remove these dissolved solids. The filters can contain carbon, which absorbs impurities; ion exchange resins, which swap metal ions for hydrogen ions, and nanoparticles of silver, to kill any microbes in the water.

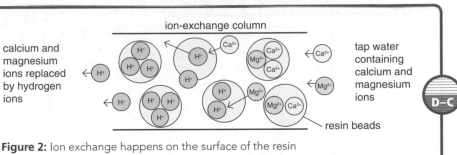

Figure 2: Ion exchange happens on the surface of the resin

D–C

Desalination

- Distillation is a form of desalination. The impure water is boiled and turned to steam before being condensed back to water. Boiling the water requires lots of energy, so it is very expensive.

How Science Works

- Whenever you consider a chemical process, one of the most important factors of the process is how much energy is needed for the process. The less energy, the cheaper the process. The Haber Process only has a yield of 15% to keep energy costs to a minimum.

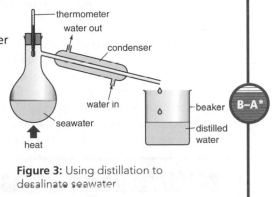

Figure 3: Using distillation to desalinate seawater

B–A*

Energy from reactions

Measuring energy transfers

- If you need to measure the energy released by burning a food or fuel, the heat can be captured by water and the temperature rise calculated.

- If you need to measure the energy released by a reaction where one or both reactants are liquid, you can do it in an insulated container then measure the temperature change.

- To calculate the energy released by a reaction use the equation $Q = m \times c \times \Delta T$ where Q is energy, m is mass of water, c is the specific heat capacity of water (4.2 J/g °C).

- ΔT is change in temperature.

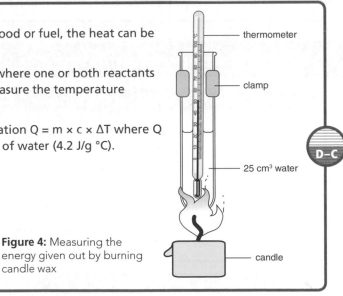

Figure 4: Measuring the energy given out by burning candle wax

D–C

Energy values

- Energy values are often quoted for foods in calories. To convert calories to joules, simply multiply the calories by 4.2.

B–A*

Improve your grade

Calculate the energy released when ethanol is burnt and heats up 250 cm³ of water by 6 °C. Assume that 1 cm³ of water weighs 1 g. **AO2 (2 marks)**

Energy from bonds

Bond breaking and forming

- In reactions, the bonds between the reactants are broken and new bonds are formed in the products. The same quantity of energy is not stored in the new bonds as in the old bonds.

- In exothermic reactions, less energy is needed in the product bonds than in the reactants and the spare energy is released as heat. In endothermic reactions, more energy is needed in the products bonds and the extra energy is taken from the surroundings, making the temperature fall. These changes can be shown as energy level diagrams.

- The energy needed to break a bond depends on the two atoms held together by the bond, and whether it is a single, double or triple bond. To find out whether it is an exothermic reaction, add up the number of each type of bond, multiply by their bond energies and add together for the reactants; repeat this for the products; then subtract the products' quantity from the reactants' quantity. If the answer is negative, then the reaction is exothermic.

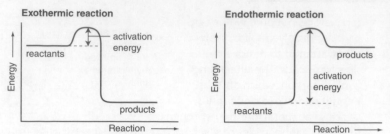

Figure 1: Energy level diagrams for an exothermic and an endothermic reaction. The bump on each diagram shows the activation energy

EXAM TIP

Calculating the energy change that happens when breaking or making bonds is a bit like doing formula mass calculations: how many bonds and what type, times the bond energy for each bond, equals the energy to break or make the bonds.

Average bond energies

- In methane (CH_4), with four C–H bonds, the first of the four is easy to break but then each next bond needs more energy to break, until the fourth one is extremely strong. The bond energy used is the average of all the four bonds, so the final calculated figure is as accurate as possible for all C–H bonds.

Remember!

Always subtract the product figure from the reactants in energy calculations. That way if you need more energy (endothermic reaction) the answer is positive or plus energy, and if you have extra energy (exothermic reaction) the answer is negative or minus energy.

Energy saving chemistry

Hydrogen, a fuel for the future

- Hydrogen gas is an alternative fuel for powering cars and heating homes. It is not a fossil fuel and can be made by the electrolysis of water.

- Hydrogen has no carbon dioxide or sulfur dioxide emissions, and releases three times as much energy as petrol does per kilogram. Unfortunately it has to be stored as a compressed gas, and needs a much larger fuel tank to travel as far as a petrol-fuelled car can.

- Burning hydrogen in a normal car engine is only 20% efficient, but if used in a fuel cell to make electricity and then power electric cars, it is 60% efficient.

How does a fuel cell work?

- Oxygen gas from the air is fed to a cathode, and hydrogen gas to an anode. A platinum catalyst removes the electrons from the hydrogen atoms, sends them round a circuit to the cathode to meet the hydrogen ions and the oxygen to react and form water.

Improve your grade

Draw a molecule of ethane (C_2H_6). Work out the number and type of bonds present. Calculate the energy stored in the bonds of ethane using these bond energy values:
C–H = 412 kJ/mol, C–C = 346 kJ/mol **AO2 (3 marks)**

Analysis – metal ions

Forming a precipitate

- Adding sodium hydroxide solution to a solution of metal ions results in the formation of the metal ion's hydroxide as a precipitate. In the case of transition metals, the hydroxides are brightly coloured and so the colour can identify the metal ion. Copper produces a blue precipitate; iron(II) produces a green precipitate; and iron(III) gives a brown one.

- Flame tests can be used to identify Group 1 and 2 metal ions, such as sodium and calcium. Table 1 shows the colours of flame tests for Groups 1 and 2.

Metal ion	Colour of flame
lithium	crimson
sodium	yellow
potassium	lilac
calcium	red
barium	green

Table 1: Flame test colours for metal ions of Groups 1 and 2

D–C

Sophisticated flame tests

- A flame photometer is an instrumental method where a metal ion is burnt and the light intensity is measured as well as its wavelength (colour). This enables the metal ion to be identified and the amount of the metal ion present to be measured.

Remember!

Group 1 metal hydroxides all dissolve in water, Group 2 produce a white precipitate, so use a flame test to identify them. Aluminium hydroxide is white, but will re-dissolve if you add more sodium hydroxide.

B–A*

Analysis – non-metal ions

Reactions in the tests

- Adding silver nitrate solution is used as a test for chloride, bromide and iodide ions. The precipitates made are coloured: silver chloride is white, silver bromide is cream, and silver iodide is yellow.

- Carbonate compounds always release carbon dioxide when tested with a little hydrochloric acid. Carbon dioxide is identified by bubbling through limewater as a white precipitate appears in the limewater.

- Sulfates are identified by adding barium chloride solution. This produces a white precipitate of barium sulfate.

D–C

Toxic but useful

- Barium salts are toxic. Barium sulfate is used to X-ray your intestines should you need it. The barium sulfate is insoluble, and doesn't react with your digestive juices, so that it just passes through your system. The barium sulfate allows the soft tissue of the intestine to become clearly visible on the X-ray.

B–A*

Improve your grade

You suspect that a solution you have been given contains some calcium chloride. Describe the tests you could use to show that calcium and chloride ions are present. **AO3 (4 marks)**

Measuring in moles

Converting mass to moles

D–C

- Chemists use moles to measure the number of atoms or molecules. One mole contains Avogadro's number (6.02×10^{23}) of particles.
- To find the number of moles, use this equation:

    moles = mass in g / relative formula or atomic mass

- To find the mass when given moles, use:

    mass in grams = moles × relative formula or atomic mass

> **Remember!**
> One mole is either the relative atomic mass or relative formula mass, in grams, of an element or compound.

Concentration

B–A*

- When making solutions, it is important to know the concentration of the solution. This is measured in moles per cubic decimetre. This way chemists know how many particles there are in the solution. A 1 mole per cubic decimetre (mol/dm^3) solution has 1 relative formula or atomic mass in grams dissolved in it.

> **Remember!**
> A cubic decimetre is also known as a litre or $1000\ cm^3$. So $500\ cm^3$ of a $1\ mol/dm^3$ solution has only 0.5 of the relative formula or atomic mass of particles; and $100\ cm^3$ has 0.1 of the A_r or M_r present.

Analysis – acids and alkalis

Calculations

D–C

- Titrations are used to measure the volumes of acid and alkali solutions that react with each other. They can be used to find the concentration of one of the solutions.
- By knowing the concentration of the acid and the volume of acid needed to neutralise the alkali, you can work out the concentration of the alkali.
- The concentration of hydrochloric acid is $0.5\ mol/dm^3$, and if $20\ cm^3$ of acid neutralised $25\ cm^3$ of sodium hydroxide solution, then the balanced equation will tell us the reacting ratio of acid to alkali:

    $NaOH(aq) + HCl(aq) \rightarrow NaCl(aq) + H_2O(\ell)$

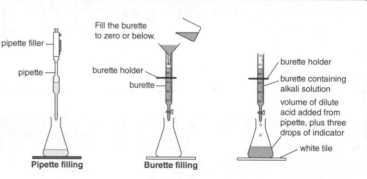

Figure 1: Titration to measure the volume of sodium hydroxide solution that reacts with a fixed volume of hydrochloric acid

- The acid and alkali react with a 1:1 ratio.
- So moles of acid used = concentration × volume (in dm^3) = $0.5 \times 20/1000 = 0.01$ moles
- Concentration of sodium hydroxide = moles / volume (in dm^3). The moles will be 0.01 as the reacting ratio is 1:1, so $0.01/(25/1000) = 0.4\ mol/dm^3$.

Using equations

B–A*

- In the reaction above, it is now possible to calculate the mass of sodium chloride produced. As 0.01 moles of hydrochloric acid was used, this means that 0.01 moles of sodium chloride would be made. So 0.01 times the relative formula mass of sodium chloride will give the answer, that is, 0.585 g.

> **EXAM TIP**
>
> Make sure you always show the working out in calculations. If you know the method, but get the arithmetic wrong, you'll lose just 1 mark; but no working and a wrong answer always means no marks.

◉ Improve your grade

Calculate the moles for the following masses of these substances.
a   11.6 g of NaCl
b   9.4 g of $MgCl_2$
c   4.44 g of $Mg(NO_3)_2$
[relative atomic masses: Na = 23, Mg = 24, O = 16, Cl = 35, N = 14] **AO2 (6 marks)**

Dynamic equilibrium

Changing the position of equilibrium

- Dynamic equilibriums are special reversible reactions. At the start, the reactants react together producing products and the products change back to reactants. After a while the proportions of products and reactant stabilise so that the forward and backward reactions are going at the same rate, so no more product is made.

- To increase or decrease the proportions of product to reactant, the conditions must be changed. Whatever the change, the equilibrium changes (shifts to a new equilibrium) to reduce the effect: this is Le Chatelier's principle.

- If the forward reaction is endothermic, then increasing the temperature increases the yield of the product. Cooling will reduce the yield. For exothermic reactions, increasing the temperature causes the forward reaction to reduce, so more reactants are made.

D–C

Changing pressure

- When making ammonia, increasing the pressure increases the forward reaction.

$$N_2(g) + 3H_2(g) \rightleftharpoons 2NH_3(g)$$

- Le Chatelier's principle operates to reduce the number of molecules present. This is done by making more ammonia (reducing 4 molecules on the left to 2 molecules on the right).

Remember!
In a dynamic equilibrium the reaction never stops, it simply continues, but the forward rate of reaction equals the backward rate of reaction.

B–A*

Making ammonia

The yield

- Ammonia is made by the Haber Process and is a dynamic equilibrium. To increase the yield, the conditions that can be changed are temperature and pressure.

- As the forward reaction is exothermic, increasing the temperature increases the backward reaction. Less is made.

- Increasing pressure increases forward reaction, reducing the molecule numbers. More is made.

- The catalyst only speeds up the reaction; it doesn't change the equilibrium position.

- The temperature (450 °C) and pressure (200 atmospheres) makes sure a small amount of ammonia is made quickly. The unreacted nitrogen and hydrogen are recycled so all the gases become ammonia eventually.

D–C

Making the hydrogen

- Methane is converted to carbon monoxide and hydrogen using steam reforming. The carbon monoxide is then reacted with water to make even more hydrogen, using a catalyst. Both reactions are equilibria and so the conditions chosen produce a quick and economic amount of hydrogen.

B–A*

Improve your grade

Look at this dynamic equilibrium. The reaction is endothermic:
$$CH_4(g) + H_2O(g) \rightleftharpoons CO(g) + 3H_2(g)$$
Describe how the equilibrium will be altered by:
a   increasing the temperature.
b   decreasing the pressure. **AO2 (2 marks)**

Alcohols

Reactions and uses of alcohols

D–C

- Alcohols dissolve in water to give neutral solutions. The molecules do not form ions.

- Alcohols such as ethanol can react with sodium metal to produce sodium ethanoate and hydrogen.

- $2Na(s) + 2C_2H_5OH(\ell) \rightarrow 2C_2H_5ONa(aq) + H_2(g)$

- Alcohols such as methanol react with oxygen by burning to make carbon dioxide. The larger the alcohol molecule, the more oxygen is needed.

$$2CH_3OH(\ell) + 3O_2(g) \rightarrow 2CO_2(g) + 4H_2O(\ell)$$

Wine into vinegar

B–A*

- Vinegar is a mixture of ethanoic acid and water. It is made by the reaction of ethanol – from wine, beer or cider – with oxygen.

$$C_2H_5OH(aq) + O_2(g) \rightarrow CH_3COOH(aq) + H_2O(\ell)$$

Carboxylic acids

Reactions

D–C

- Ethanoic acid is part of a homologous series of compounds called carboxylic acids. They all contain the functional group –COOH.

- They dissolve in water to produce hydrogen ions, so they are acidic. The –COOH group becomes –COO⁻(aq) and H^+(aq).

- Carboxylic acids react with metal carbonates to produce carbon dioxide, water and the metal salt of the acid, for example sodium ethanoate.

- Alcohols react with carboxylic acids to produce smelly compounds called esters. Esters are used to make perfumes and flavour foods.

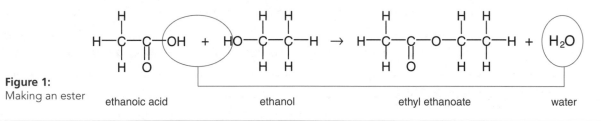

Figure 1:
Making an ester   ethanoic acid   ethanol   ethyl ethanoate   water

Strong and weak acids

B–A*

- When a carboxylic acid dissolves in water some, but not all, of the molecules change into ions. This means that they are weak acids. They do not have a pH value of 1, like hydrochloric acid which is a strong acid, but a pH of 3 or 4 which makes them weak acids.

Improve your grade

Write a chemical equation to describe how ethanoic acid ionises in water to form a weak acid. Explain why ethanoic is a weak acid. **AO1 (3 marks)**

C3 Summary

The periodic table

The modern periodic table can be seen as an arrangement of the elements in terms of their electronic structures. The trends in reactivity within groups in the periodic table can be explained by electronic structures.

The elements in Group 1, the alkali metals, have a low density; and react with non-metals to form ionic compounds in which the metal ion carries a charge of +1.

Compared with the elements in Group 1, Transition elements have higher melting points and higher densities, and have coloured compounds. They are useful as catalysts.

The elements in Group 7, the halogens, react with metals to form ionic compounds in which the halide ion carries a charge of -1.

Water

Soft water readily forms lather with soap, hard water does not. Hard water contains dissolved compounds, usually of calcium or magnesium.

Water filters containing carbon, silver and ion-exchange resins can remove some dissolved substances from tap water to improve the taste and quality. Chlorine and fluorides may be added to drinking water for health reasons.

Permanent hard water can be made soft by removing the dissolved calcium and magnesium ions by adding sodium carbonate, or using ion exchange.

Permanent hard water remains hard when it is boiled. Temporary hard water is softened by boiling.

Energy changes

The energy produced by a reaction can be calculated from the measured temperature change.

Simple energy-level diagrams can be used to show the relative energies of reactants and products, the activation energy and the overall energy change of a reaction.

The relative amounts of energy released when substances burn can be measured by simple calorimetry using $Q = mc \Delta T$.

Hydrogen can be burned as a fuel in combustion engines. It can also be used in fuel cells that produce electricity to power vehicles.

During a chemical reaction, energy must be supplied to break bonds, energy is released when bonds are formed.

Analysis and quantitative chemistry

Flame tests can be used to identify metal ions. Adding sodium hydroxide to transition metal ions produces a coloured precipitate.

Non-metal ions can be identified using chemical tests.

The volumes of acid and alkali solutions that react with each other can be measured by titration using a suitable indicator.

If the concentration of one of the reactants is known, the results of a titration can be used to find the concentration of the other reactant.

Ammonia

In gaseous reactions, an increase in pressure will favour the reaction that produces the least number of molecules as shown by the symbol equation for that reaction.

The raw materials for the Haber process to make ammonia are nitrogen and hydrogen. The conditions for the reaction are important.

Alcohols, carboxylic acids and esters

Alcohols contain the functional group –OH. Ethanol can be oxidised to ethanoic acid, either by chemical oxidising agents or by microbial action.

Ethanoic acid is a member of the carboxylic acids, which have the functional group –COOH. Carboxylic acids dissolve in water to produce weak acidic solutions.

Ethyl ethanoate is the ester produced from ethanol and ethanoic acid. Esters have the functional group –COO– They are volatile compounds with distinctive smells and are used as flavourings and perfumes.

C1 Improve your grade

Page 6

Draw the electronic structures of the following atoms with proton numbers 3, 9, 11, 16, and 20.
AO2 (5 marks)

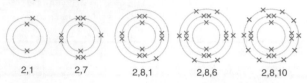

2,1   2,7   2,8,1   2,8,6   2,8,10

The answer gets 4 marks (grade C) as the first four are correct. In the last one, where the proton number is 20, the answer has the last 10 electrons in the third orbit rather than six in the third and two in the fourth orbits. For full marks the structure should be 2,8,8,2.

Page 7

Explain why oxygen has a molecule with a double covalent bond and fluorine has only a single covalent bond. You should refer to the number of outer electrons in both atoms in your answer. **AO2 (4 marks)**

Fluorine has electronic structure 2,7, so needs one electron to fill the shell. So it makes a single bond. Oxygen is 2,6, and needs two electrons to fill the shell, so it makes two bonds.

This answer gets 2 marks (grade D). It correctly gives the electronic structure of the two atoms, but fails to mention sharing electrons with the other atom, or the need to achieve a Noble gas electronic structure.

Page 8

Balance these two chemical equations:

a $Mg(s) + O_2(g) \rightarrow MgO(s)$
b $C_3H_8(g) + O_2(g) \rightarrow CO_2(g) + H_2O(\ell)$

AO2 (2 marks)

a $2Mg(s) + O_2(g) \rightarrow 2MgO(s)$
b $C_3H_8(g) + O_2(g) \rightarrow 3CO_2(g) + 4H_2O(\ell)$

This answer gets 1 mark (grade C). Part a) is correctly balanced, but in part b) the hydrogen and carbons are correct, but the oxygens have not been adjusted. It should read $5O_2$.

Page 9

Draw a table to show which of these metals is extracted by carbon and which by electrolysis. **AO1 (2 marks)**

iron, magnesium, aluminium, copper, zinc, lead, potassium, and calcium

carbon	electrolysis
iron	magnesium
zinc	aluminium
lead	potassium
copper	calcium

This answer gets 2 marks (grade C). All the metals are correctly placed. You should remember that copper is extracted by carbon, but is often purified by electrolysis.

Page 10

Explain the difference between oxidation and reduction. **AO1 (2 marks)**

Oxidation is gaining oxygen and reduction is the loss of oxygen.

This answer gets 2 marks (grade C). Oxidation and reduction are the opposite of each other. In a reaction, if one substance is oxidised the other is reduced.

Page 11

Explain why the waste heaps of an old copper mine would be a suitable site for phytomining or bioleaching. **AO2 (2 marks)**

Copper mines produce waste which still has copper ore in it.

This answer gets 1 mark (grade C). It recognises that the waste heaps contain copper ore, but fails to explain that phytomining or bioleaching would be a cheap and effective way of obtaining the copper in the waste.

Page 12

Explain why titanium and aluminium metals are both reactive and corrosion free. **AO1 (2 marks)**

Both metals get coated with an oxide layer very quickly in the air. This layer seals the surface, and prevents further corrosion. The layer is only two or three molecules thick.

This answer gets 2 marks (grade B). It clearly makes the two points needed, that an oxide layer is formed, and that this prevents further corrosion.

Page 13

Suggest why using biofuels may create more problems than it solves. **AO2 (3 marks)**

Biofuels are grown on land that can be used to grow food. This may reduce the amount of food the world produces, and lead to starvation. It may also cause food prices to rise beyond what people can afford.

This answer gets 3 marks (grade A). It clearly makes the point about land use, food costs, and food production. It doesn't mention that rainforests may be felled in order to grow profitable crops to make biofuel from.*

Page 14

Draw out the structure of the straight chained alkanes with 5 carbon atoms and 7 carbon atoms.
AO2 (2 marks)

This answer gets 1 mark (grade C). The second structure should have seven carbon atoms, not six. Check your diagrams carefully.

Page 15

Draw structural diagrams to show how $C_{12}H_{26}$ can be converted to C_3H_6, and another molecule. State which of the two products is unsaturated, and why. **AO1 and 2 (3 marks)**

It is the C_3H_6 that is unsaturated because it has a double bond.

This answer gets 2 marks (grade B). The diagram for the conversion is good, but on the C_3H_6 diagram there is a missing hydrogen on the left. Always count that each carbon atom has 4 bonds. The answer recognises propene as the unsaturated molecule and correctly gives the reason.

Page 16

PVC (polyvinylchloride) is a commonly used polymer. Its monomer has the structure:

Show, using three monomer molecules, the structure and bonding of a PVC polymer chain. **AO2 (2 marks)**

Each of the lines represents a single covalent bond or a pair of shared electrons.

This answer gets 2 marks (grade B). The diagram shows the structure and bonding correctly. The additional statement makes it clear that the candidate knows that this molecule has covalent bonding, and what covalent bonding is.

Page 17

Describe the difference between recycling polymer waste and re-using it by incineration. Give two reasons why recycling is more environmentally friendly than incineration. **AO1 (3 marks)**

Recycling polymer waste means re-using it as a polymer after some processing. Re-using it by incineration is when the polymer is burnt. The polymer is changed to heat and carbon dioxide.

This answer gets 1 mark (grade C). There is a clear description of the two processes but the second part of the question is ignored in the answer. Recycling is less damaging as it needs less energy for recycling than for making new polymers, saves on using fossil fuels, and stops polluting gases caused by incineration; it also reduces waste sent to landfill.

Page 18

Explain why biofuels are considered to be carbon neutral. **AO1 (2 marks)**

Biofuels are carbon neutral because they only put into the atmosphere the same carbon dioxide as they took out by growing.

This answer gets 2 marks (grade C). The answer scores full marks, but could be improved by adding that photosynthesis takes the carbon dioxide out of the air, rather than using the word 'growing'.

Page 19

Draw a diagram to show how an emulsifier would surround a water droplet to make a water in oil emulsion. **AO2 (2 marks)**

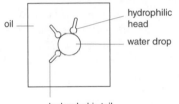

This answer gets 2 marks (grade A/B). The diagram is clear and shows both hydrophobic and hydrophilic ends of the emulsifier. It clearly shows that it is a drop of water in oil, and not the other way round. It doesn't label the emulsifier.

Page 20

Explain how the Atlantic Ocean is getting wider each year. Suggest another part of the world where a similar process may be taking place. **AO1 (3 marks)**

America is drifting away from Britain by convection currents in the sea. This is also happening with Europe.

This answer gets 0 marks (grade D). The candidate knows about convection currents, but has placed them in the sea, rather than in the mantle. The increasing gap is caused by the tectonic plates drifting on the mantle-driven convection currents and, where the two plates are moving apart, fresh rock is being formed. The middle of the Pacific Ocean is another place where this is happening.

Page 21

Sketch a timeline showing how the proportions of water vapour, carbon dioxide, oxygen, and nitrogen have changed since the Earth was formed.
AO2 (4 marks)

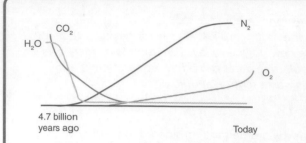

This answer gets 4 marks (grade A). It clearly shows how each of the four gases' concentration in the air has changed over time. It would have been helpful, but not necessary, to add in when life began (3.5 billion years ago), and also when life began on the land.

Page 22

Explain why the Miller–Urey experiment does not prove how life evolved. **AO3 (2 marks)**

It shows how amino acids, which are the building blocks of life, were formed.

This answer gets 1 mark (grade C). The answer is not an explanation, it simply states the outcome of the experiment. It should have mentioned that amino acid formation was necessary for life to develop, but that this is only a theory, as none of the amino acids formed proteins or were alive at all.

C2 Improve your grade

Page 24

Work out the relative atomic mass (A_r) of chlorine. Chlorine has two isotopes, 25 per cent is chlorine-37, 75 per cent is chlorine-35. **AO2 (2 marks)**

25 x 37 = 925, 75 x 35 = 2625 so 3550/100 = 35.5

This answer gets 2 marks (grade B). The mean value is correctly calculated.

Page 25

Use the periodic table to:

a Work out the electronic structure of these elements: magnesium, chlorine, nitrogen, aluminium, calcium. **AO2 (5 marks)**

b Work out how many outer electrons each of these elements has: strontium, arsenic, gallium, astatine, germanium. **AO2 (5 marks)**

a Mg = 2,8,2 Cl = 2,8,7 N = 2,7
  Al = 2,8,3 Ca = 2,8,8,2

b Sr 2, As 5, Ga 1, At 7, Ge 2.

This answer gets 7 marks (grade B). For part a) all except N are correct. The problem is that the first orbit has not been subtracted from the 7, it should be 2,5. For part b) the candidate has correctly found three, but for Ga and Ge has used the periodic table but only counted in from the transition metals instead of using the group number.

Page 26

Draw diagrams to show how calcium and fluorine lose and gain electrons to make the compound calcium fluoride (CaF_2). **AO1 (3 marks)**

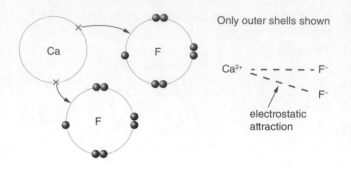

This answer gets 3 marks (grade A). To save time, the candidate has only drawn the outer electron shells. The answer clearly shows what happens to the electrons, states correctly the two ions formed, and gives the ionic bond as an electrostatic attraction.

Page 27

Explain in detail why sodium chloride:

a has a high melting point
b conducts electricity when in solution.
  AO1 (3 marks)

a Sodium chloride has a crystal lattice so the melting point is high.

b It conducts electricity along the lattice.

This answer gets 0 marks (grade E). The answer recognises the crystal lattice, but should explain that this forms a giant structure where six ions are attracted equally to six others, making a very strong structure that needs lots of energy to break (melt). Part b) fails to mention that the ions are now free to move and can carry the electric charge/current through the solution.

Page 28

Draw diagrams to show the covalent bonds present in these compounds:
HCl, H_2, CO_2, NH_3, and CH_4. **AO1 (5 marks)**

$$H \overset{\times}{\underset{\bullet}{}} \quad Cl$$

This answer gets 4 marks (grade B). Clear dot-and-cross diagrams are used; but for ammonia, the pair of unbonded electrons has been missed out.

Page 29

Draw diagrams to show the difference between the structure of a thermosetting and a thermosoftening polymer. **AO1 (2 marks)**

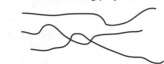

thermosetting polymer   thermosoftening polymer

This answer gets 2 marks (grade C). Both chain structures have been clearly drawn and labelled.

Page 30

Explain how the metal lattice allows for metals to be stretched without breaking. **AO1 (2 marks)**

The atoms can be easily pulled slightly apart from each other.

This answer gets 0 marks (grade E). The atoms slide past each other distorting the original shape, but not separating out.

Page 31

Look at Figure 6. It shows the retention–time graph for a gas chromatogram sample.

a   How many different components were in the sample?
b   List the components in order of the quantity of each present. Start with the most plentiful. **AO3 (2 marks)**

a   there are four substances

b   A B C D E is the solvent.

This answer gets 0 mark (grade E). Because a) there are five peaks so there are five substances and b) the answer suggests that the candidate thinks E is the solvent. The graph has been wrongly read: the height of the peaks is the important feature. So A C D B E is the correct answer.

Page 32

Calculate the formula of lithium oxide made from 5.6 g of lithium and 6.4 g of oxygen. **AO2 (2 marks)**

Li 5.6/7 = 0.8, O = 6.4/16 = 0.4 ratio is 8:4 or 2:1, so Li_2O

This answer gets 2 marks (grade A). It is a very difficult calculation that only the most able students can get right. The method clearly shows how to get the answer.

Page 33

Calculate the mass of calcium oxide that can be made from 100 tonnes of calcium carbonate.

The balanced equation for the reaction is:

$CaCO_3(s) \rightarrow CaO(s) + CO_2(g)$ **AO2 (3 marks)**

M_r $CaCO_3$ = 40 + 12 + 16 + 16 + 16 = 100

M_r CaO = 40 + 16 = 56, from equation 1 mole $CaCO_3$ produces 1 mole CaO, so 100 tonnes makes 56 tonnes.

This answer gets 3 marks (grade B). A well-presented calculation that shows how to work out M_r, and use the equation to calculate the moles, and then the masses involved.

Page 34

Use Figure 2 to calculate the rate of the reaction at the following times:

a   0 to 0.5 minutes

b   1.0 to 2.0 minutes

c   2.5 to 3.0 minutes. **AO3 (3 marks)**

a   6.4 in 0.5 min, so in 1 min 12.8

b   changes from 8.8 to 11.0 in the minute so rate is 2.2

c   changes from 12.0 to 12.4 in the minute so rate is 0.4

This answer gets 3 marks (grade B). The answers to a) and b) are correct, but have no units. In this case the units are cm^3 per minute or cm^3/min. For part c) the answer is wrong, the value is only for 0.5 minutes, so must be multiplied by 2 to get the right answer, which needs units.

Page 35

A student was investigating a chemical reaction. She decided to increase the temperature of the reaction, whilst decreasing the concentration of one reactant. Use collision theory to explain why the rate of reaction remained unchanged. **AO3 (2 marks)**

The increased temperature has no effect as she only changed one of the reactant's concentration not both. To change the rate, she needs to change both the reactants' concentration.

This answer gets 0 marks (grade E). Increasing the temperature does increase the rate, but reducing the concentration of one reactant will counteract the change. The increase in reacting collisions caused by the temperature rise is equalled by the reduction in reacting collisions caused by less of the reactant being present.

Page 36

Use Figure 1 to find the rate of the reaction when the concentration is 2 mol/dm³ after 1 minute. **AO2 (2 marks)**

this is 6

This answer gets 0 marks (grade D). The answer has been simply read off the graph, and has no units. A tangent should have been drawn to enable the gradient of the curve to be found, so that the rate at 1 minute could be found. The x-axis value would be 1.8, y-axis would be 4, so $4/1.8 = 2.2$ cm^3 per minute.

Page 37

What is activation energy?

Explain how a catalyst lowers the activation energy for a particular reaction. **AO1 (3 marks)**

Activation energy is energy needed for a reaction to start.

A catalyst works by providing an easier path for the reaction.

This answer gets 2 marks (grade C). The first part is just worth a mark. More detail about reacting collisions would have made the answer better. For the second part, the second mark could be gained if there was more detail about it being an energetically lower reaction path.

Page 38

Sketch an energy-level diagram for the exothermic reaction that takes place when nitric acid reacts with sodium hydroxide. Write on the line for 'reactants', the names of the reactants; and on the line for 'products', the products. **AO1 (4 marks)**

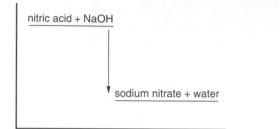

nitric acid + NaOH

sodium nitrate + water

This answer gets 4 marks (grade B). The levels are correctly drawn and the reactants and products all named. NaOH, the formula, is acceptable when asked for a name but if asked for a formula, the name would be marked as wrong.

Page 39

Describe the method to make some solid chromium chloride from chromium metal. **AO3 (4 marks)**

Use hydrochloric acid, add some powdered chromium metal to it, when it stops reacting filter it, then evaporate the solution.

This answer gets 2 marks (grade C). The basic steps are there, but you need to check the solution made is neutral with, for example, pH paper. Warming the mixture would help to ensure complete reaction. More detail is needed about the evaporation technique, including the need to partially evaporate and then to filter it to obtain some crystals to dry.

Page 40

Write half-equations to show the reactions, at both the anode and cathode, for the electrolysis of copper chloride ($CuCl_2$). **AO2 (2 marks)**

$Cu^{2+} + e^- \rightarrow Cu$

$Cl^{2-} + e^- \rightarrow Cl_2$

This answer gets 0 marks (grade D). The answer does not show any understanding. Copper ion is correctly described, but that is all. The need to balance the electrons and charges is attempted, but only as far as wrongly giving chloride ions a 2^- charge. Correct equations are $Cu^{2+} + 2e^- \rightarrow Cu$ and $2Cl^- + 2e^- \rightarrow Cl_2$

C3 Improve your grade

Pages 42

Describe the two assumptions Mendeleev made when designing his periodic table. How did these assumptions lead to his table being accepted?
AO1 (4 marks)

He left spaces for elements that he didn't know and made predictions about them. When the elements were discovered and their properties found to be the same as his predictions this convinced people he was right. He did this for gallium (eka-aluminium).

This answer gets 2 marks (grade C). The student has given one point, about leaving spaces in the periodic table, and explains clearly how this assumption helped to convince people the table was useful. The answer should also refer to changing the order to reflect the chemistry as he did with Te and I (tellurium and iodine), which also helped convince people.

Page 43

Explain why potassium is a more reactive metal than lithium. **AO1 (2 marks)**

The outer electron in potassium is easier to remove than in lithium because it is further away from the nucleus. This means it is less strongly held as there are more electron shells between the nucleus and the outer electron, and so potassium will react more violently than lithium.

This answer gets 2 marks (grade A). The answer clearly states that both the shielding effect and the distance from the nucleus affect the ease of losing the outer electron.

Page 44

Explain why permanent hard water cannot be softened by boiling yet temporary hard water can.
AO2 (3 marks)

Temporary hard water contains calcium hydrogencarbonate that is turned into calcium carbonate by heating. Permanent hard water contains other calcium compounds that are not affected by heating.

This answer gets 2 marks (grade C). It correctly states that temporary hard water contains calcium hydrogencarbonate that turns into calcium carbonate, and that permanent hard water is not affected by boiling. The last mark would be gained by stating that calcium carbonate is insoluble, so its formation removes the calcium ions.

Page 45

Calculate the energy released when ethanol is burnt and heats up 250 cm³ of water by 6 °C. Assume that 1 cm³ of water weighs 1 g. **AO2 (2 marks)**

250 g × 4.2 J/g/°C × 6 = 6300J

This answer gets 2 marks (grade C). It gains 1 mark for correct substitution of values and 1 mark for giving the answer with its units.

Page 46

Draw a molecule of ethane (C_2H_6). Work out the number and type of bonds present. Calculate the energy stored in the bonds of ethane using these bond energy values:

C–H = 412 kJ/mol, C–C = 346 kJ/mol

AO2 (3 marks)

6 C–H bonds, 2 C–C bonds.

so (412×6) + (346×2) = 3164 kJ/mol

This answer gets 2 marks (B grade). The diagram is correct, and it is clear that the student knows how to do the calculation, but they have counted the C–C bond twice, once at each carbon atom, so the final answer is wrong, it should be 2818 kJ/mol.

Page 47

You suspect that a solution you have been given contains some calcium chloride. Describe the tests you could use to show that calcium and chloride ions are present. **AO3 (4 marks)**

Use a flame test to show calcium is present. It should give a red colour. Chloride is tested for using nitric acid, and silver nitrate solutions. This gives a white precipitate which if left in light will turn grey as it reacts with light.

This answer gets 4 marks (grade A). It is a clear answer, giving both test and result for each ion.

Page 48

Calculate the moles for the following masses of these substances.

a   11.6 g of NaCl

b   9.4 g of $MgCl_2$

c   4.44 g of $Mg(NO_3)_2$

[atomic masses: Na = 23, Mg = 24, O = 16, Cl = 35, N = 14]

AO2 (6 marks)

a   11.6/(23 + 35) = 0.2

b   9.4/(24 + 35 + 35) = 0.1

c   4.44/24+(14+(16 x 6) = 0.033

This answer gets 5 marks (grade B) The first two answers are correct and gain two marks each, the last answer incorrectly calculated the relative formula mass of $Mg(NO_3)_2$ by using only one N atom instead of two. The correct answer should be 0.03.

Page 49

Look at this dynamic equilibrium. The reaction is endothermic:

$CH_4(g) + H_2O(g) \rightleftharpoons CO(g) + 3H_2(g)$

Describe how the equilibrium will be altered by:

a   increasing the temperature.

b   decreasing the pressure.

AO2 (2 marks)

a   As the reaction is endothermic, the reaction will go to the right or more products will be made.

b   Decreasing the pressure will favour an increase in the number of molecules. More products will be made.

This answer gets 2 marks (grade A). The student knows the effect of changing both variables on the equilibrium. In both cases, the forward reaction is favoured.

Page 50

Write a chemical equation to describe how ethanoic acid ionises in water to form a weak acid.

Explain why ethanoic is a weak acid. **AO1 (3 marks)**

Weak acids are not fully ionised, but strong acids are. If you have one hundred acid molecules in a strong acid all of them ionise to form hydrogen ions. In a weak acid only about twenty or thirty of the acid molecules ionise so there are less hydrogen ions, so the acid is weak.

This answer gets 3 marks (grade A). The student clearly understands the differences between strong and weak acids.

Understanding the scientific process

As part of your assessment, you will need to show that you have an understanding of the scientific process – How Science Works.

This involves examining how scientific data is collected and analysed. You will need to evaluate the data by providing evidence to test ideas and develop theories. Some explanations are developed using scientific theories, models and ideas. You should be aware that there are some questions that science cannot answer and some that science cannot address.

Collecting and evaluating data

You should be able to devise a plan that will answer a scientific question or solve a scientific problem. In doing so, you will need to collect data from both primary and secondary sources. Primary data will come from your own findings – often from an experimental procedure or investigation. While working with primary data, you will need to show that you can work safely and accurately, not only on your own but also with others.

Secondary data is found by research, often using ICT – but do not forget books, journals, magazines and newspapers are also sources. The data you collect will need to be evaluated for its validity and reliability as evidence.

Presenting information

You should be able to present your information in an appropriate, scientific manner. This may involve the use of mathematical language as well as using the correct scientific terminology and conventions. You should be able to develop an argument and come to a conclusion based on recall and analysis of scientific information. It is important to use both quantitative and qualitative arguments.

Changing ideas and explanations

Many of today's scientific and technological developments have both benefits and risks. The decisions that scientists make will almost certainly raise ethical, environmental, social or economic questions. Scientific ideas and explanations change as time passes and the standards and values of society change. It is the job of scientists to validate these changing ideas.

How science ideas change

From the information you have learnt, you will know that science is a process of developing, then testing theories and models. Scientists have been carrying out this work for many centuries and it is the results of their ideas and trials that has provided us with the knowledge we have today.

However, in the process of developing this knowledge, many ideas were put forward that seem quite absurd to us today, such as this example from the 17th Century.

> In 1667, Johann Joachim Becher published Physical Education, in which he set out the basis of what was to become the Phlogiston Theory. The theory itself was formally stated in 1703 by Georg Ernst Stahl, a German professor of Medicine and Chemistry. He suggested that in all flammable substances there is something called 'phlogiston', a substance without colour, odour, taste, or mass that is given off during burning. 'Phlogisticated' substances are those that contain phlogiston and, on being burned, are 'dephlogisticated'. The ash of the burned material is thought to be the true material. The theory was widely supported for much of the eighteenth century. Joseph Priestley, who is credited with the discovery of oxygen, defended the theory but it was eventually disproved by Antoine Lavoisier.

Reliability of information

It is important to be able to spot when data or information is presented accurately, and just because you see something online or in a newspaper does not mean that it is accurate or true.

Think about what is wrong in this example, based on a document from a government official to a parliamentary committee. Look at the answer at the bottom of the page to check that your observations are correct.

RECENT DERAILMENTS IN THE STATE

Note that this timeline is for background information only, and is not meant to be a comprehensive analysis of rail safety incidents.

Aug. 5, 2005: 9 cars of a 144-car train derail spilling highly-acidic caustic soda into the Yukon river; a preliminary government investigation states that federal safety regulations were broken by the railroad company...

Answer
Caustic soda is highly alkaline, not highly acidic.

The glossary contains terms useful for your revision. Page numbers are given for items that are covered in this book.

coal 9   solid fossil fuel formed from plant material – composed mainly of carbon

collision frequency 35–7   number of collisions per second between the particles involved in a chemical reaction

collision theory 35–6, 41   relates reaction rates to the frequency and energy of collisions between the reacting particles

combustion 13   process where substances react with oxygen, releasing heat

compound 6, 27, 41   substance composed of two or more elements joined together by chemical bonds, for example, H_2O

concentration 32, 35–6, 47   amount of chemical present in a given volume of a solution – usually measured as g/dm^3 or mol/dm^3

concrete 9, 23   mixture of cement, sand, aggregate and water

condensation 14   change of state when a substance changes from a gas or vapour to a liquid: the substance condenses

conduction (electrical) 26, 30   flow of electrons through a solid, or ions through a liquid

conductor 40   material that transfers energy easily

continental drift 20   movement of continents relative to each other

continental plate 21   tectonic plate carrying large landmass, though not necessarily a whole continent

convection current 20   when particles in a liquid or gas gain energy from a warmer region and move into a cooler region, being replaced by cooler liquid or gas

convection 23   heat transfer in a liquid or gas – when particles in a warmer region gain energy and move into cooler regions carrying this energy with them

core (of Earth) 20, 23   layer in centre of Earth, consisting of a solid inner core and molten outer core

covalent bond 7, 14, 23, 28–29, 40   bond between atoms in which some of their outer electrons are shared

cracking 15–16, 23   oil refinery process that breaks down large hydrocarbon molecules into smaller ones

crust 20, 23   surface layer of Earth made of tectonic plates

crystallise 9, 41   form crystals from a liquid – for example, by partly evaporating a solution and leaving to cool

current 30   flow of electricity round a circuit – carried by electrons through solids and by ions through liquids

D

decay (biological) 22   the breakdown of organic material by microorganisms

delocalised electrons 30   electrons not attached to any particular atom, so free to move through the structure, allowing electrical conduction – present in metals and graphite

direct current (d.c.) 40   electric current where the direction of the flow of current stays constant, as in cells and batteries

distillation 45   process for separating liquids by boiling them, then condensing the vapours

DNA 21–22   deoxyribonucleic acid – the chemical from which chromosomes are made: its sequence determines genetic characteristics, such as eye colour

double covalent bond 7, 15–16, 46   two covalent bonds between the same pair of atoms – each atom shares two of its own electrons plus two from the other atom

E

earthquake 21   shaking and vibration at the surface of the Earth resulting from underground movement or from volcanic activity

electrode 11   solid electrical conductor through which current passes into and out of liquid during electrolysis – and at which the electrolysis reactions take place

electrolysis 9, 11, 23, 27, 31, 40–41   decomposing an ionic compound by passing a d.c. electric current through it while molten or in solution

electrolyte 40   solution or molten substance that conducts electricity

electron 6–7, 23–25, 41–43, 51   small particle within an atom that orbits the nucleus (it has a negative charge)

electronic configuration 6, 42–43, 51   the arrangement of electrons in shells, or energy levels, in an atom

electronic structure (or configuration) 25, 27–8, 43   arrangement of electrons in shells, or energy levels, in an atom

element 6–7, 24–25, 42, 51   substance made out of only one type of atom

empirical formula 32   ratio of elements in a compound, as determined by analysis – for example, CH_2O for glucose (molecular formula $C_6H_{12}O_6$)

emulsifier 19, 23   a substance that prevents an emulsion from separating back into oil and water

emulsion 19, 23   a thick, creamy liquid made by thoroughly mixing an oil with water (or an aqueous solution)

endothermic reaction 38, 41, 46   chemical reaction which takes in heat, or energy from other sources

energy levels 38, 46   electrons in shells around the nucleus – the further from the nucleus, the higher the electron's energy level

energy output 38, 45   the energy transferred away from a device or appliance – it can be either useful or wasted

energy 38, 45   the ability to 'do work'

enzyme 37   biological catalyst that increases the speed of a chemical reaction but is not used up in the process

essential oils 18   oils found in flowers, giving them their scent – they vaporise more easily than natural oils from seeds, nuts and fruit

ethanol 13, 16, 23, 50   an alcohol that can be made from sugar and used as a fuel

exothermic reaction 37–38, 41, 46   chemical reaction which gives out heat

F

fermentation 13, 16, 23   process in which yeast converts sugar into ethanol (alcohol)

flammable 14   catches fire and burns easily

formula (for a chemical compound) 25 group of chemical symbols and numbers, showing which elements, and how many atoms of each, a compound is made up of

forward reaction 33–34, 38, 41, 49 reaction from left to right in an equation for a reversible reaction

fossil 24 preserved remains of a long-dead organism

fossil fuel 13, 22 fuel such as coal, oil or natural gas, formed millions of years ago from dead plants and animals

fractional distillation 14, 16, 21, 23 process that separates the hydrocarbons in crude oil according to size of molecules

fractionating column 14 tall tower in which fractional distillation is carried out at an oil refinery

fractions 23 the different substances collected during fractional distillation of crude oil

fuel cell 46, 51 device that generates electricity directly from a fuel, such as hydrogen, without burning it

fullerenes 30 cage-like carbon molecules containing many carbon atoms, for example, C_{60}, buckminsterfullerene

G

gas chromatography 31 method that uses a gas to carry the substances through a long, thin tube of adsorbent

giant covalent structure 29 solid structure made up of a regular arrangement of covalently bonded atoms – may be an element or a compound

giant ionic structure 41 solid structure made up of a regular arrangement of ions in rows and layers

global dimming 13 gradual decrease in the average amount of sunlight reaching Earth's surface

global warming 13 gradual increase in the average temperature of Earth's surface

greenhouse gas 22 a gas such as carbon dioxide that reduces the amount of heat escaping from Earth into space, thereby contributing to global warming

group 6, 42–4 within the periodic table the vertical columns are called groups

H

Haber process 45, 49 industrial process for making ammonia

halogens 27, 39, 44 reactive non-metals in Group 7 of the periodic table

hard water 39, 44 water supply containing dissolved calcium or magnesium salts – these react with soap, making it hard to form a lather

hydrocarbon 14, 23, 32 compound containing only carbon and hydrogen

hydrophilic 19, 23 water-loving (attracted to water, but not to oil) – used to describe parts of a molecule

hydrophobic 9, 23 water-fearing (attracted to oil, but not to water) – opposite of hydrophilic

hypothesis 20 an idea that explains a set of facts or observations – a basis for possible experiments

I

immiscible 19 liquids that do not mix, but form separate layers, are immiscible

insoluble 39 not soluble in water (forms a precipitate)

insoluble salt 39, 41 salt which is not soluble in water, so forms a precipitate

intermolecular forces 28–29 forces between molecules

ion 23, 40–41, 47 atom (or group of atoms) with a positive or negative charge, caused by losing or gaining electrons

ionic bonding 7, 23, 26 chemical bond formed by attractions between ions of opposite charges

ionic compound 41 compound composed of positive and negative ions held together in a regular lattice by ionic bonding, for example, sodium chloride

isotopes 24 forms of element where their atoms have the same number of protons but different numbers of neutrons

J

joule 45 unit used to measure energy

L

lattice 27, 29–30, 41 regular arrangement of ions or atoms in a solid – may be covalent or ionic

lava 21 magma that has erupted onto the surface of Earth

limestone 8–9, 23 type of rock consisting mainly of calcium carbonate

limewater 9, 47 calcium hydroxide solution

lithosphere 20 the rocky, outer section of the Earth, consisting of the crust and upper part of the mantle

low-grade ore 11 ore containing only a small percentage of metal

M

magma 20–1 molten rock found below Earth's surface

mantle 20, 23 semi-liquid layer of the Earth beneath the crust

mass number 6, 23 total number of protons and neutrons in the nucleus of an atom – always a whole number

mass spectrometer 31 instrument for identifying chemicals by measuring their relative formula mass very accurately

mass 31 a measure of the amount of 'stuff' in an object

melting 27 change of state of a substance from liquid to solid

metallic bonding 30 type of bonding in metals – a regular lattice of metal ions is held together by delocalised electrons

methane 8, 46, 49 the simplest hydrocarbon, CH_4 – main component of natural gas

mole 25, 48 unit for counting atoms and molecules – one mole of any substance contains the same number of particles

molecular ion 31 ion formed when an electron is knocked off a molecule in a mass spectrometer

molecule 28 two or more atoms held together by covalent chemical bonds

molten 9, 40 made liquid by keeping the temperature above the substance's melting point

monomers 16, 29   small molecules that become chemically bonded to each other to form a polymer chain

mortar 9, 23   mixture of cement, sand and water

N

nanometer 30   unit used to measure very small length (1 nm = 0.000 000 001 m, or one-billionth of a metre)

nanoparticles 30   very small particles (1–100 nanometres in size)

nanoscience 41   the study of nanoparticles

nanotube 30   carbon molecule in the form of a cylinder

neutralization 32, 38, 41   reaction between an acid and a base to make a salt and water (H^+ ions react with OH^- or O^{2-} ions)

neutron 6, 23–24, 41   small particle that does not have a charge – found in the nucleus of an atom

noble gas 7, 25, 42   unreactive gas in Group 0 of the periodic table

nucleus 24   central part of an atom that contains protons and neutrons

nutrient 23   substance in food that we need to eat to stay healthy, such as protein

O

oceanic plate 21   tectonic plate under the ocean floor – it does not carry a continent

OH group 50–51   an oxygen atom bonded to a hydrogen atom and found in all alcohols

oil (crude) 14–16, 23   liquid fossil fuel formed from animals and plants that lived 100 million years ago

oil (from a plant) 23   liquid fat obtained from seeds, nuts or fruit

ore 9, 23   rock from which a metal is extracted, for example iron ore

oxidation 10, 32, 40–41   process that increases the amount of oxygen in a compound – opposite of reduction

P

Pangea 20   huge landmass with all the continents joined together before they broke up and drifted apart

percentage yield 33   percentage yield = actual yield ÷ theoretical yield x 100)

period 7   horizontal row in the periodic table

periodic table 7, 42, 51   a table of all the chemical elements based on their atomic number

pH scale 49   scale from 0 to 14 which shows how acidic or alkaline a substance is

photochromic 30   photochromic materials change colour in response to changes in light level

photosynthesis 13, 37   process carried out by plants where sunlight, carbon dioxide and water are used to produce glucose and oxygen

phytomining 11–12, 23   using growing plants to absorb metal compounds from soil, burning the plants, and recovering metal from the ash

phytoremediation 12   cleaning up contaminated soil by using growing plants to absorb harmful metal compounds

pipette 48   used to measure out an exact volume of liquid

plastics 16–17   compounds produced by polymerisation, capable of being moulded into various shapes or drawn into filaments and used as textile fibres

plate boundaries 21   edges of tectonic plates, where they meet or are moving apart

pollution 17–18   presence of substances that contaminate or damage the environment

poly(ethene) 16, 23, 29   plastic polymer made from ethene gas (also called polythene)

polymer 16–17, 23, 29   large molecule made up of a chain of monomers

polymerisation 16–17, 23, 29   chemical process that combines monomers to form a polymer: this is how polythene is formed

precipitate 39, 41, 47   solid product formed by reacting two solutions

precipitation 32, 39   reaction between two solutions to form a solid product (a precipitate)

products 46, 49   chemicals produced at the end of a chemical reaction

proton 6, 23–24   small positive particle found in the nucleus of an atom

Q

quarry 8, 23   place where stone is dug out of the ground

R

radioisotope 24   a radioactive isotope of an element

reactants 32–7, 41, 46, 49   chemicals that are reacting together in a chemical reaction

reaction conditions 44, 47   physical conditions under which a reaction is performed, for example, temperature and pressure

reaction rate (average) 34   total amount of reaction ÷ total time

reaction rate (initial) 32, 34   reaction rate at the start of the reaction

reaction rate 34–6, 38, 41   the speed at which a chemical reaction takes place – measured as the amount of reaction per unit time

reactivity series 9, 26   list of metals in order of their reactivity with oxygen, water and acids

reduction 9–10, 31, 40   process that reduces the amount of oxygen in a compound, or removes all the oxygen from it – opposite of oxidation

relative atomic mass 34, 31, 41, 47–48   average mass of all the atoms in an element, taking into account the presence of different isotopes – often rounded to the nearest whole number

relative formula mass 25, 32, 41, 47–48   total mass of all atoms in a formula = each relative atomic mass × number of atoms present

relative molecular mass 31   same as relative formula mass, but limited to elements or compounds that have separate molecules

renewable resource 16   energy resource that is constantly available or can be replaced as it is used

retention factor (R$_f$) 31 used to help identify individual spots in a chromatogram (R$_f$ = distance moved by the spot ÷ distance moved by the solvent)

retention time (R$_t$) 31 time taken for a component to travel through the tube of adsorbent during gas chromatography

reverse reaction 34 reaction from right to left in the equation for a reversible reaction

reversible reaction 33–34, 38, 41, 49 a reaction that can also occur in the opposite direction – that is, the products can react to form the original reactants again

rutile 12 an ore of titanium – impure titanium oxide (TiO_2)

S

salt 38–39, 41 compound composed of metal ions and non-metal ions – formed by acid–base neutralisation

saturated fat 15 solid fat, most often of animal origin, containing no C=C double bonds

saturated hydrocarbon 14–15, 19 hydrocarbon containing only single covalent bonds

shape memory alloy 10, 30 alloy that 'remembers' its original shape and returns to it when heated

shells 6–7, 23, 25 electrons are arranged in shells (or orbits) around the nucleus of an atom – also known as 'energy levels'

slag 9 waste material produced during smelting of a metal – it contains unwanted impurities from the ore

smart material 10, 17, 30 material which changes in response to changes in its surroundings, such as light levels or temperature

smelting 9, 12 extracting metal from an ore by reduction with carbon – heating the ore and carbon in a furnace

soften (water) 39, 44 treat water so as to remove the calcium ions that cause hardness

soluble salt 39, 41 salt which dissolves in water

solvent 31 liquid in which solutes dissolve to form a solution

stainless steel 11 steel alloy containing chromium and nickel to resist corrosion

steam cracking 15 cracking hydrocarbons by mixing with steam and heating

steam distillation 18, 45, 51 process of blowing steam through a mixture to vaporise volatile substances – used to extract essential oils from flowers

steel 10–11 alloy of iron and steel, with other metals added depending on its intended use

sub-atomic particle 6 particle that makes up an atom, such as proton, neutron or electron

subduction zone 21 area of ocean floor in which an oceanic plate is sinking beneath a continental plate

successful collisions 35 collisions with enough energy to break bonds in the reactant particles, and thus cause a reaction

sugar 13, 16, 22, 37 sweet-tasting compound of carbon, hydrogen and oxygen such as glucose or sucrose

sulfur dioxide 12, 18 poisonous, acidic gas formed when sulfur or a sulfur compound is burned

surface area (of a solid reactant) 35–6 measure of the area of an object that is in direct contact with its surroundings

symbol (for an element) 6 one or two letters used to represent a chemical element, for example C for carbon or Na for sodium

T

tectonic plate 20–1, 23 section of Earth's crust that floats on the mantle and slowly moves across the surface

theoretical yield 33 mass of product that a given mass of reactant should produce according to calculations from the equation – the actual yield is always less than this

thermal decomposition 15, 23 chemical reaction in which a substance is broken down into simpler chemicals by heating it

thermochromic 30 thermochromic materials change colour in response to changes in temperature

thermosetting polymer 29, 41 plastic polymer that sets hard when heated and moulded for the first time – it will not soften or melt when heated again

thermosoftening polymer 29, 41 plastic polymer that softens and melts when heated and reheated

thin layer chromatography (TLC) 31 chromatography using a plate coated with a thin layer of powdered adsorbent

titration 47 procedure to determine the volume of one solution needed to react with a known volume of another solution

tracer 47 radioactive element used to track the movement of materials, such as water through a pipe or blood through organs of the body

transition metals 37, 43–47 group of metal elements in the middle block of the periodic table – includes many common metals

triple covalent bond 46 three covalent bonds between the same two atoms – each atom shares three of its own electrons plus three form another atom

U

unsaturated fats 15, 19, 23 liquid fats, containing C=C double bonds – usually from plants or fish

unsaturated hydrocarbon 15 hydrocarbon containing one or more C=C double bonds

V

vent 21 crack or weak spot in the Earth's crust, through which magma reaches the surface

volcano 20–1 landform (often a mountain) where molten rock erupts onto the surface of the planet

W

water of crystallisation 34 water molecules present in crystals of some metal salts – shown separately in the formula, for example, hydrated copper sulfate, $CuSO_4.5H_2O$

Y

yeast 16, 22 single-celled fungus used in making bread and beer

yield 49 mass of product made from a chemical reaction

Workbook

NEW GCSE SCIENCE

Chemistry

for AQA A Higher

Author: Rob Wensley

Revision guide +
Exam practice workbook

The key to successful revision is finding the method that suits you best. There is no right or wrong way to do it.

Before you begin, it is important to plan your revision carefully. If you have allocated enough time in advance, you can walk into the exam with confidence, knowing that you are fully prepared.

Start well before the date of the exam, not the day before!

It is worth preparing a revision timetable and trying to stick to it. Use it during the lead up to the exams and between each exam. Make sure you plan some time off too.

Different people revise in different ways and you will soon discover what works best for you.

Some general points to think about when revising

- Find a quiet and comfortable space at home where you won't be disturbed. You will find you achieve more if the room is ventilated and has plenty of light.

- Take regular breaks. Some evidence suggests that revision is most effective when tackled in 30 to 40 minute slots. If you get bogged down at any point, take a break and go back to it later when you are feeling fresh. Try not to revise when you're feeling tired. If you do feel tired, take a break.

- Use your school notes, textbook and this Revision guide.

- Spend some time working through past papers to familiarise yourself with the exam format.

- Produce your own summaries of each module and then look at the summaries in this Revision guide at the end of each module.

- Draw mind maps covering the key information on each topic or module.

- Review the Grade booster checklists on pages 114–116.

- Set up revision cards containing condensed versions of your notes.

- Prioritise your revision of topics. You may want to leave more time to revise the topics you find most difficult.

Workbook

The Workbook allows you to work at your own pace on some typical exam-style questions. You will find that the actual GCSE questions are more likely to test knowledge and understanding across topics. However, the aim of the Revision guide and Workbook is to guide you through each topic so that you can identify your areas of strength and weakness.

The Workbook also contains example questions that require longer answers (**Extended response questions**). You will find one question that is similar to these in each section of your written exam papers. The quality of your written communication will be assessed when you answer these questions in the exam, so practise writing longer answers, using sentences. The **Answers** to all the questions in the Workbook are detachable for flexible practice and can be found on pages 121–134.

At the end of the Workbook there is a series of **Revision checklists** that you can use to tick off the topics when you are confident about them and understand certain key ideas.

Remember

There is a difference between learning and revising.

When you revise, you are looking again at something you have already learned. Revising is a process that helps you to remember this information more clearly.

Learning is about finding out and understanding new information.

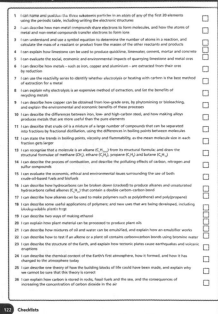

Atoms, elements and compounds

1 Malachite is an ore of copper. It contains both rock, and copper carbonate. When heated with carbon it produces copper, and carbon dioxide.

(a) Name the element obtained from malachite.

_____ [1 mark]

(b) Name the useful compound in malachite.

_____ [1 mark]

(c) Explain why you know this substance is a compound.

_____ [1 mark]

(d) Explain why malachite is a mixture.

_____

_____ [2 marks]

D–C

Inside the atom

1 Phosphorus is an element in the periodic table.

(a) Use information in the box to help you complete these sentences.
An atom of phosphorus contains _____ protons, _____ electrons and has a mass of _____.
It has _____ neutrons.

[4 marks]

(b) Name or give the symbol of an element that:

(i) has the same number of outer electrons as phosphorus.

_____

(ii) has two more outer electrons than phosphorus.

_____

(iii) has one less proton than phosphorus.

_____ [3 marks]

(c) Draw a diagram
to show the electronic
structure of phosphorus.

[2 marks]

31

P

15

D–C

2 Sodium and potassium both react with water to produce hydrogen gas and an alkali.

(a) Draw a diagram
to show the electronic
structure of potassium.

[2 marks]

(b) Draw a diagram
to show the electronic
structure of sodium.

[2 marks]

(c) Explain why both sodium and potassium have very similar reactions with water.

_____ [2 marks]

B–A*

Element patterns

1 (a) How are the elements arranged in the periodic table?

_____ [1 mark]

(b) What are horizontal rows of the periodic table called?

_____ [1 mark]

(c) What are vertical columns of the periodic table called?

_____ [1 mark]

(d) Elements in the same vertical column are sometimes called 'families'. Explain why.

_____ [2 marks]

D–C

2 Elements of Group 0 are known as the noble gases.

(a) Name one noble gas, and give its electronic structure.

_____ [2 marks]

B–A*

(b) Explain why noble gases form very few compounds.

_____ [2 marks]

Combining atoms

1 This is a space filling model of a water molecule (H_2O). The atoms are joined together by covalent bonds.

(a) Describe a covalent bond.

_____ [1 mark]

(b) Draw a dot and cross diagram to show the electron arrangement in the water molecule.

[2 marks]

Sodium chloride (NaCl) is a compound where two different elements are present. The atoms transfer electrons to each other.

(c) Give the electronic configuration of a sodium atom.

_____ [1 mark]

(d) Describe how the sodium atom becomes a positively charged ion.

_____ [2 marks]

D–C

(e) Explain how this helps the sodium to form a compound with the chlorine.

_____ [3 marks]

2 Here is a diagram of a carbon dioxide molecule. O=C=O

(a) Explain what the = symbol means.

_____ [1 mark]

B–A*

(b) Draw a dot and cross diagram to show the covalent bonding in carbon dioxide.

[2 marks]

Chemical equations

1 Copper carbonate can be used to make copper. When heated with carbon it produces copper and carbon dioxide.

D–C

   (a) Complete this word equation

   copper + carbon → copper + _____   [1 mark]
   carbonate

   (b) This is the balanced symbol equation for the reaction

   $2CuCO_3 + C \rightarrow 2Cu + 3CO_2$

   (i) How many molecules of carbon dioxide are produced in the reaction? _____ [1 mark]

   (ii) How many atoms of copper can be made for each atom of reacting carbon? _____ [1 mark]

2 Copper carbonate has many reactions. Here are unbalanced symbol equations for two of the reactions. Balance each equation by inserting numbers in front of each substance if necessary to balance them.

B–A*

   (a) $CuCO_3 + HCl \rightarrow CuCl_2 + CO_2 + H_2O$ [1 mark]

   (b) $CuCO_3 + NaOH \rightarrow Cu(OH)_2 + Na_2CO_3$ [1 mark]

Building with limestone

1 A company obtains large quantities of limestone by quarrying.
These quarries are often in beautiful parts of the country such as the Peak District.

The company wants to open a new quarry. Describe the advantages and disadvantages, socially and economically for the local population in opening the new quarry.

D–C

   _____

   _____

   _____

   _____

   _____ [4 marks]

2 Limestone rock areas often have many caves and potholes. These are often caused by the effect of rainwater.

   (a) Describe how the rainwater dissolves the limestone.

   _____

   _____

B–A*

   _____ [3 marks]

   (b) In limestone caves you often find stalactites and stalagmites, which are made from solid calcium carbonate. Explain how these form.

   _____

   _____ [2 marks]

Heating limestone

1 Heating limestone converts it into calcium hydroxide. Calcium hydroxide dissolves in water to make limewater.

(a) Limewater can be used to test for a gas in the air. Which gas is this? _____ [1 mark]

(b) Describe how to use limewater to test for this gas, and what you would see if the gas were present.

_____ [2 marks]

(c) Cement is made from limestone.

(i) What substance is limestone heated with to make cement? _____ [1 mark]

(ii) Cement is used to make concrete and mortar. What is the difference between mortar and concrete?

_____ [2 marks]

(d) Mortar holds bricks together, concrete can be used to make beams for use in constructing houses. Explain why these two materials made from limestone have these different uses.

_____ [2 marks]

D–C

---

2 (a) Long beams made from concrete often crack in the middle. Describe how beam manufacturers can prevent beams cracking in use.

_____ [2 marks]

(b) Many people say that concrete dries when it becomes solid. Explain why this statement is wrong.

_____ [2 marks]

force needed to break beam in kN x 1000 vs percentage of gravel in the concrete

(c) This graph shows how the strength of concrete changes according to the volume of sand and gravel in the concrete. The mass of cement used remained the same throughout the investigation.

(i) Describe the trend of the graph from 0% to 50% gravel in the concrete. _____ [1 mark]

(ii) Suggest why the concrete strength drops rapidly at 75% gravel in the mixture.

_____ [2 marks]

B–A*

Metals from ores

1 Metals such as iron, copper and zinc are obtained from their ores. Here is a table of some common ores.

(a) Name the ore from the table that can be used to obtain

name of ore	formula
haematite	Fe_2O_3
bauxite	Al_2O_3
litharge	PbO
zincite	ZnO

(i) iron _____ [1 mark]

(ii) zinc _____ [1 mark]

(iii) aluminium _____ [1 mark]

(b) Which ore in the table cannot be obtained by reduction of the ore with carbon. Explain your answer. _____ [2 marks]

(c) What is meant by reduction? _____ [1 mark]

D–C

---

2 Aluminium, sodium and potassium are all more plentiful in the Earth's crust than iron. In the 1850s aluminium was a very rare and expensive metal, and cost more than gold. Today aluminium is still more expensive than iron, but it is cheap enough to be used to make lemonade cans.

(a) How is aluminium extracted from its ore today? _____ [1 mark]

(b) Explain why this method of reduction is more expensive than reduction with carbon.

_____ [2 marks]

(c) Suggest why aluminium was more expensive in the 1850s. _____ [1 mark]

B–A*

Extracting iron

1 Iron ore (Fe_2O_3) is reduced inside a blast furnace to make iron.

(a) Name the two other substances added to the blast furnace.

(i) _____ **(ii)** _____ [2 marks]

(b) Hot air is blown through the blast furnace. This produces large volumes of carbon dioxide gas. Explain why.

_____ [3 marks]

(c) Slag is formed in the process. Describe how slag is formed.

_____ [1 mark]

D–C

2 Iron is produced in a blast furnace by reduction. This chemical equation summarises the reaction.

$$Fe_2O_3(s) + ...CO(g) \rightarrow ...Fe(\ell) + ...CO_2(g)$$

(a) Balance the equation. _____ [1 mark]

(b) What is the name of $CO(g)$? _____ [1 mark]

(c) Explain how the $CO(g)$ has been produced. _____

_____ [2 marks]

(d) Write the formula of the substance that has been reduced. _____ [1 mark]

(e) Write the formula of the substance that has been oxidised. _____ [1 mark]

B–A*

Metals are useful

1 The diagram represents the structure of a pure metal. Use the diagram to help you answer these questions.

(a) Describe why a metal can be bent without breaking.

_____ [2 marks]

(b) Explain how a metal can conduct electricity.

_____ [2 marks]

(c) Describe how a metal is different from an alloy.

_____ [2 marks]

(d) Explain why an alloy is harder than the original metal.

_____ [2 marks]

D–C

2 Nitinol is a nickel titanium memory or smart alloy.

(a) Alloys are often used when each metal they are made from would be unsuitable for that use. Explain why.

_____ [2 marks]

(b) Nitinol is a memory alloy. What does this mean?

_____ [1 mark]

(c) Strips of smart alloy can be cooled, stretched and then used to hold the parts of a broken bone together. Evaluate the benefits of using a smart alloy to hold broken pieces of a single bone together whilst it heals.

_____ [3 marks]

B–A*

Iron and steel

1 Stainless steel can be made by adding chromium and nickel to iron to make an alloy we call stainless steel.

(a) Give two advantages of using stainless steel instead of iron to make cutlery.

_____

_____ [2 marks]

(b) Vanadium can also be added to iron to make a different stainless steel. Suggest why having different types of stainless steel is useful.

_____ [1 mark]

(c) Which part of the periodic table are all these metals found in?

_____ [1 mark]

D–C

Copper

1 Copper is purified using electrolysis of copper sulfate solution.
This diagram shows the process.

(a) What is the anode made from?

_____ [1 mark]

(b) What is the cathode made from?

_____ [1 mark]

(c) Where does the pure copper collect?

_____ [1 mark]

(d) Where do the impurities collect?

_____ [1 mark]

(e) Describe what happens to copper ions at the cathode.

_____ [2 marks]

D–C

2 Copper ores have very low percentages of copper in them. At some old copper mines, the heaps of waste rock contain more copper than the remaining deposits.

(a) Describe how growing plants on the waste rock heaps can be used to obtain copper ores.

_____

_____ [3 marks]

(b) Describe how introducing copper tolerant bacteria can be used to obtain the copper from the rock.

_____

_____ [3 marks]

(c) Although both processes take a long time, suggest why both are now economic to do.

_____ [2 marks]

B–A*

Aluminium and titanium

1 Aluminium is extracted from its ore by electrolysis.

D–C

   (a) What is meant by electrolysis? _____ [1 mark]

   (b) Name the main aluminium ore. _____ [1 mark]

   (c) **(i)** Why is aluminium an expensive metal?

       _____ [2 marks]

      (ii) Explain how dissolving the aluminium ore in cryolite reduces the cost of aluminium manufacture.

       _____ [3 marks]

2 This equation represents the reduction of aluminium oxide to aluminium metal. This chemical equation summarises the reaction.

$$...Al_2O_3 \rightarrow ...Al + ...\ O_2$$

B–A*

   (a) Balance the equation. [1 mark]

   (b) Titanium is produced by reacting with magnesium. Suggest why titanium is a very expensive metal to make.

      _____ [1 mark]

   (c) Both titanium and aluminium are high in the reactivity series, but do not react with acids or alkalis. Explain why.

      _____ [2 marks]

   (d) Suggest why titanium is a suitable material to use to make artificial joints in the body.

      _____ [2 marks]

Metals and the environment

1 Many metals are recycled.

D–C

   (a) Evaluate the economic benefits of recycling metals against using new metal. Your answer should suggest two advantages and two disadvantages of recycling metals.

      _____ [6 marks]

   (b) Describe the environmental disadvantages of mining metal ores.

      _____ [3 marks]

2 Phytomining is a new method of obtaining metal ores from *brownfield* sites. This flow chart shows the process.

land contaminated with toxic metal compounds	→	land used to grow plants	→	plants harvested to become metal ores	→	land fit to build homes on

B–A*

   (a) What is a *brownfield* site? _____ [1 mark]

   (b) Explain how the plants remove the metal compounds from the ground.

      _____ [2 marks]

   (c) Describe how the harvested plants are turned into metal ore.

      _____ [2 marks]

   (d) What is the social benefit of making the land fit for people to live on?

      _____ [2 marks]

A burning problem

1 In 2010, an Icelandic volcano erupted sending large quantities of ash into the atmosphere, and gases such as sulfur dioxide and carbon dioxide. Jet aircraft were prevented from flying over a large part of Europe.

(a) Explain how the ash may cause global dimming.

_____ [2 marks]

(b) Describe how the volcanic eruption may lead to more acid rainfall.

_____ [2 marks]

(c) Suggest how the eruption could lead to global warming.

_____ [1 mark]

D–C

2 Burning dry plant material such as wood in a garden bonfire produces carbon dioxide, lots of ash and a little pure carbon called charcoal.

(a) Explain why some of the carbon in the wood becomes charcoal.

_____ [2 marks]

(b) In a cigarette the same burning process produces another compound called carbon monoxide instead of carbon dioxide.

(i) Write the chemical formula of carbon dioxide. _____ [1 mark]

(ii) Explain how carbon monoxide is different from carbon dioxide.

_____ [1 mark]

(iii) Describe how carbon monoxide is a health risk to smokers.

_____ [2 marks]

B–A*

Reducing air pollution

1 Car exhaust systems are fitted with catalytic converters.

(a) Why are cars fitted with a catalytic converter?

_____ [1 mark]

(b) What environmental benefit is gained by using catalytic converters.

_____ [1 mark]

(c) To reduce vehicle emissions, some cars are designed to use ethanol instead of petrol as a fuel.

Explain why ethanol is a green fuel. _____ [3 marks]

D–C

2 Biofuels are now added to petrol and diesel. Using these fuels made from plants is reducing our dependency on non-renewable resources. Biofuels are considered to be carbon neutral and do not contribute to global warming.

(a) Explain why biofuels are considered to be carbon neutral.

_____ [2 marks]

(b) Describe why biofuels are classed as renewable fuels.

_____ [2 marks]

B–A*

(c) Land that used to grow food crops is now growing crops for biofuels. Describe the problems this may cause now and in the future. _____ [2 marks]

Crude oil

1 Crude oil is a mixture of many different hydrocarbons.

(a) What is meant by *hydrocarbon*?

_____ [1 mark]

(b) Here is a diagram of how the crude oil is separated into more useful substances.

(i) Name the separation process. _____ [1 mark]

(ii) Name the substance with the highest boiling point. _____ [1 mark]

(iii) Name the substance with the lowest boiling point. _____ [1 mark]

(c) Explain how the process separates crude oil into different useful substances.

_____ [4 marks]

2 The boiling points of several of the substances shown in the diagram are given in this table, with the mean number of carbon atoms in the substance.

(a) Why is the mean number of carbon atoms given for these substances?

_____

[1 mark]

(b) Describe the trend shown by the data.

_____

[1 mark]

substance	mean number of C atoms	boiling point in °C
diesel	14	250
kerosene	12	180
naphtha	10	110
petroleum gas	8	20

(c) Explain why the boiling point is related to the mean number of carbon atoms in the mixture.

_____ [2 marks]

Alkanes

1 Here is the general formula for an alkane. C_nH_{2n+2}

(a) Explain how you could use this to work out the formula of octane which has 8 carbon atoms.

_____ [2 marks]

(b) As alkanes have the same basic structure, they have similar properties that change as the molecule gets bigger. Explain how each of these properties changes as the alkane molecule gets bigger:

(i) flammability? _____

(ii) boiling point? _____

(iii) viscosity? _____ [3 marks]

2 (a) Describe the type of bonding in alkanes.

_____ [2 marks]

(b) Methane (CH_4) and ethane (C_2H_6) are the first two alkanes. Draw their structures.

_____ [2 marks]

(c) Explain why alkanes have similar chemical properties that change as the molecules get larger.

_____ [2 marks]

Cracking

1 Cracking hydrocarbon molecules is a very important part of an oil refinery's work. The longer alkane molecules are broken down into shorter molecules.

(a) Write the formula of decane.

_____ [1 mark]

(b) Name molecule X.

_____ [1 mark]

octane

molecule X

decane

(c) What does the = symbol mean in the structural diagram of molecule X? _____ [1 mark]

(d) Why is molecule X a very useful chemical?

_____ [2 marks]

D–C

2 In cracking, the large alkane molecule is broken down into two smaller molecules by thermal decomposition.

(a) Explain what is meant by thermal decomposition.

_____ [1 mark]

(b) Steam cracking uses steam and a temperature of 850 °C, and catalytic cracking uses a catalyst and a temperature of 600 °C. Describe one advantage and one disadvantage of each type of cracking.

_____ [4 marks]

B–A*

Alkenes

1 Ethene is the simplest alkene

(a) How are alkenes different from alkanes?

_____ [1 mark]

(b) Use the diagram to explain why alkenes are more reactive than alkanes.

_____ [2 marks]

(c) Alkenes are said to be unsaturated hydrocarbons. What do chemists mean by unsaturated?

_____ [1 mark]

D–C

2 A student wanted to check if a hydrocarbon had any double bonds present. The student put 1 cm³ of the hydrocarbon in a test tube.

(a) What should the student add to the hydrocarbon to test whether any double bonds are present?

_____ [1 mark]

(b) What would the student expect to see?

_____ [2 marks]

(c) How would this be different for a hydrocarbon with no double bonds?

_____ [2 marks]

B–A*

Making ethanol

1 Ethanol (C_2H_5OH) can be made by two methods: fermentation from sugars, or from ethene (C_2H_4) obtained from crude oil.

(a) What is added to ethene to produce ethanol? _____ [1 mark]

D–C

(b) Ethanol made from sugars is a renewable biofuel. What is a biofuel?

_____ [1 mark]

(c) Suggest two advantages and two disadvantages of using sugar to produce ethanol for use as a car fuel.

_____ [4 marks]

---

2 Ballpoint-ink stains on clothing can be difficult to remove. One washing powder manufacturer suggests that the affected area should be rubbed with a solvent other than water before washing.

(a) Name a different solvent from water.

B–A*

_____ [1 mark]

(b) Explain why ballpoint-ink stains may be removed by this solvent but water cannot remove them.

_____ [2 marks]

Polymers from alkenes

1 Alkenes such as ethene can be used to make polymers such as poly(ethene).

(a) What is meant by a polymer?

_____ [1 mark]

D–C

(b) What word do we use to describe the small molecule used to make the polymer?

_____ [1 mark]

(c) Explain how the ethene molecules join together to make poly(ethene).

_____ [2 marks]

---

2 Thirty years ago thick glass bottles were used to transport dangerous chemicals. This type of bottle is made from poly(ethene) and is now used to transport dangerous chemicals.

(a) Suggest two reasons why the polymer bottle has replaced glass bottles for transporting dangerous chemicals.

_____ [2 marks]

(b) There are many different types of man-made polymers. Guttering is made from PVC, which is made from this small molecule:

B–A*

(i) Explain why we need different polymers.

_____ [1 mark]

(ii) Compare how PVC guttering's properties need to be different from wrapping polymers such as clingfilm.

_____ [2 marks]

Designer polymers

1 Plastic polymers are easily moulded into shape, and are low density (lightweight), waterproof and resistant to acids and alkalis. Increasingly, car and aircraft bodies are being made from designer plastic polymers such as carbon fibre-reinforced plastic.

  (a) Give three reasons why these expensive plastics are being used in cars and aircraft.

  _____ [3 marks]

  (b) Gore-tex® is a designer fabric, it lets water vapour through but not liquid water. Explain why it is a good choice from which to make an outdoor coat.

  _____ [2 marks]

  (c) Suggest why the use of designer polymers is likely to increase in the future.

  _____ [1 marks]

D–C

2 Plastic carrier bags from supermarkets are a source of pollution. They do not decompose, and can hurt or kill wildlife that get entangled in them. Recently, cornstarch-polymer carrier bags are being produced that are biodegradable.

  (a) What is meant by biodegradable? _____ [1 mark]

  (b) Suggest why cornstarch bags may be more biodegradable than plastic bags.

  _____ [2 marks]

  (c) Surgeons carrying out operations are also starting to use thread made from cornstarch polymers to stitch inside people's bodies. Explain why using biodegradable stitches during an operation is better for the patient.

  _____ [2 marks]

B–A*

Polymers and waste

1 There are three ways to dispose of polymer waste. It can be sent to landfill sites, it can be burnt in incinerators or it can be recycled.

  (a) There are problems with all three methods of disposal. Suggest a disadvantage of:

  (i) putting the polymer waste in a landfill site. _____

  (ii) incinerating the polymer waste. _____

  (iii) recycling the polymer waste. _____ [3 marks]

  (b) Evaluate which of the three methods of disposal provides the most effective use of the limited resources available to make polymers. Give reasons for your answer.

  _____ [3 marks]

D–C

2 Biodegradable polymers are often made from a mixture of ordinary hydrocarbon-based polymers and cornstarch. They are not usually recycled.

  (a) Describe what is likely to happen to a biodegradable polymer when it decomposes.

  _____ [1 mark]

  (b) Suggest why this makes it unsuitable to be used to make compost, even though it contains plant material, such as cornstarch, and is biodegradable.

  _____ [2 marks]

  (c) Compare the benefits of disposing of the biodegradable polymer by landfill and by incineration.

  _____ [2 marks]

B–A*

Oils from plants

1 Vegetable oils are oils that are made from plants such as sunflower, maize, olives, almonds and walnuts. They have many uses, such as in foodstuffs, cosmetics, and as fuels such as biodiesel.

(a) Why are plant oils suitable for use as fuels?

_____ [1 mark]

(b) From which parts of the plant is the plant oil obtained?

_____ [1 mark]

(c) Describe how the plant oil is obtained from the plant.

_____ [2 marks]

(d) Cooking food in plant oils raises the energy content of the food, and changes the flavour of the food. Explain why cooking a potato in plant oil makes the potato taste different from when it is cooked in water.

_____ [3 marks]

2 Essential oils can be obtained from flowers for perfumes using this equipment.

(a) Describe what is happening at A.

_____

_____ [2 marks]

(b) Describe the substances present in the pipe at B.

_____

_____ [2 marks]

(c) Give the scientific name for part C.

_____ [1 mark]

(d) Where has the water at D come from?

_____ [1 mark]

(e) Describe the difference between a vegetable oil and a mineral oil.

_____

_____ [2 marks]

Biofuels

1 Biofuels are fuels made from animal or plant materials. They are considered to be 'greener' than fossil fuels.

(a) Evaluate the economic and environmental benefits of using biofuels. Your answer should suggest two advantages and two disadvantages.

_____

_____

_____

_____

_____ [6 marks]

(b) Plant oils are far more viscous than diesel. Describe how they can be modified to be used in diesel engines.

_____

_____ [2 marks]

Oils and fats

1 Low-fat spreads and margarines are made from plant oils. Butter is made from milk produced by cows. Some students tested butter, hard margarine, soft margarine and low-fat spread. They tested each spread for saturated fat by dissolving a sample of each in a little ethanol, adding 2 cm³ of orange bromine water and timing how long the reaction took.

    (a) What would you expect to see happen? _____ [1 mark]

    (b) Explain why the test is not a fair one.

    _____ [2 marks]

Here are their results.

    (c) Which spread has the most saturated molecules present? Explain your answer.

    _____ [2 marks]

    (d) Explain the difference between a 'saturated' fat and an 'unsaturated' fat.

    _____ [2 marks]

Spread	Mean time for the reaction to complete in seconds
butter	2
hard margarine	4
soft margarine	7
low-fat spread	8

D–C

2 Soft or liquid vegetable fats are often converted into harder vegetable fats by reacting the fat with hydrogen using a catalyst such as nickel.

    (a) What is the purpose of using a nickel catalyst?

    _____ [1 mark]

B–A*

    (b) Draw a diagram to show how this section of a vegetable fat molecule would be hydrogenated.

```
    H H H H H
    | | | | |
H—C—C—C=C—C—
    | |     |
    H H     H
```

    [2 marks]

Emulsions

1 In 2010, there was a large oil spillage off the coast of the USA between New Orleans and Florida. The crude oil did not mix with the seawater but floated on top of it. The crude oil and seawater were mixed together using a dispersant or detergent, to make an emulsion.

    (a) What word do we use to describe two liquids that don't mix together? _____ [1 mark]

    (b) What type of substance is the dispersant or detergent? _____ [1 mark]

D–C

    (c) When making salad cream, egg yolk is added to a mixture of oil, water, vinegar and powdered mustard.

       (i) Name the aqueous substances in the mixture. _____ [1 mark]

       (ii) What is the purpose of the egg yolk? _____ [1 mark]

2 This is a diagram of a molecule that can be used as a dispersant for crude oil.

hydrophobic end   hydrophilic end

    (a) What does the term 'hydrophilic' mean? _____ [1 mark]

    (b) What does the term 'hydrophobic' mean? _____ [1 mark]

B–A*

    (c) Describe how this molecule allows crude oil to mix with seawater to form an emulsion.

    _____ [3 marks]

Earth

1 This is a diagram of the Earth's structure.

Complete the table below.

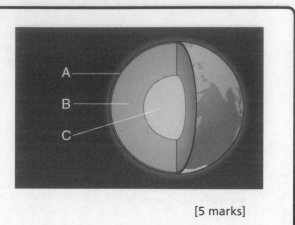

Part	Name	Its structure
A		solid rock
B		
C		

[5 marks]

2 Look at the diagram in the question above.

(a) Part B has convection currents in it. Explain how these convection currents occur.

_____ [2 marks]

(b) Part A floats on the top of part B, and is broken into small sections called tectonic plates. Explain how these plates and the convection currents form mountains.

_____ [2 marks]

(c) Rivers flow down the mountains. Describe how the rivers form valleys.

_____ [2 marks]

Continents on the move

1 Alfred Wegener suggested in 1915 that originally all the continents were joined together in one continent called Pangea. Most scientists thought he was wrong.

(a) Explain why most scientists disagreed with Wegener's theory.

_____ [2 marks]

(b) Wegener showed that South America could fit together with Africa like jigsaw pieces. Give two other pieces of evidence that were found which persuaded scientists to support his theory.

_____ [2 marks]

(c) New York is slowly moving away from London by a few centimetres each year. Describe how this is happening, using Wegener's theory.

_____ [2 marks]

2 The Mid-Atlantic Ridge is a range of underwater mountains that are forming where the North American plate meets the Eurasian plate. Iceland is an island where this range of mountains has grown so high that it is now above sea level.

Explain how this mountain ridge is formed.

_____

_____ [2 marks]

Earthquakes and volcanoes

1 This diagram shows the main tectonic plates of the Earth's crust.

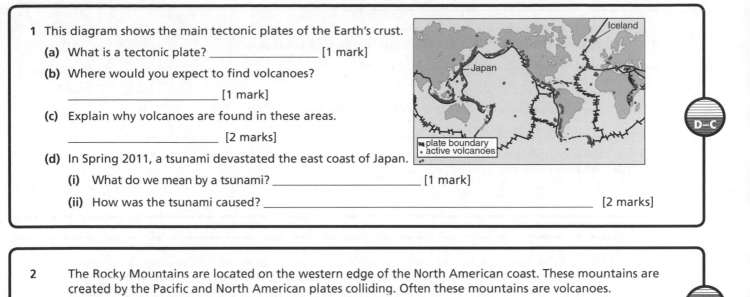

(a) What is a tectonic plate? _____ [1 mark]

(b) Where would you expect to find volcanoes?

_____ [1 mark]

(c) Explain why volcanoes are found in these areas.

_____ [2 marks]

(d) In Spring 2011, a tsunami devastated the east coast of Japan.

(i) What do we mean by a tsunami? _____ [1 mark]

(ii) How was the tsunami caused? _____ [2 marks]

D–C

2 The Rocky Mountains are located on the western edge of the North American coast. These mountains are created by the Pacific and North American plates colliding. Often these mountains are volcanoes.

Describe how the colliding plates cause mountains and volcanoes to be formed.

_____ [4 marks]

B–A*

The air we breathe

1 The composition of the air has not always been the same since the Earth formed. It has changed.

(a) Where do we think the early atmosphere came from?

_____ [1 mark]

(b) This graph shows how three gases in the air have changed since the Earth formed. Use information from the graph to help you answer these questions.

(i) How old is the Earth? _____ [1 mark]

(ii) Which letter, A to E, shows when plants were first able to photosynthesise? _____ [1 mark]

(iii) When did life start to live on the land? _____ [1 mark]

(c) Explain why life could exist only in the sea until ozone appeared in the atmosphere.

_____ [2 marks]

D–C

2 Air is a mixture of many gases. We obtain pure gases from the air by fractional distillation.

Here is some information about three important gases in the air.

(a) What is meant by 'fractional distillation'?

_____ [1 mark]

Name of gas	Boiling point in °C
nitrogen	−196
oxygen	−183
argon	−186

(b) Name two other gases in the air that must be removed before the air is cooled. _____ [2 marks]

(c) The air is cooled to −200 °C before distillation.

(i) Why is it cooled to −200 °C? _____ [1 mark]

(ii) Which of the three gases distils off first? Explain your answer.

_____ [2 marks]

B–A*

The atmosphere and life

1 Primitive life first evolved about 3.4 billion years ago. No one knows how it evolved, although there are many different theories. Most of the theories suggest that life formed in the seas.

(a) Explain why scientists do not know how life began.

_____ [1 mark]

(b) Suggest why most theories believe that life began in the seas.

_____ [1 mark]

(c) Experiments to prove how life began usually concentrate on devising a method to make amino acids. Explain why.

_____ [2 marks]

D–C

2 Miller and Urey suggested that lightning was a key variable in the formation of life. They used a solution representing seawater from 4 billion years ago, filled their apparatus with the gases present in the air 4 billion years ago and passed electric sparks through the atmosphere. They analysed the seawater after a week.

(a) Name the three gases they used to make the atmosphere of 4 billion years ago.

_____ [1 mark]

(b) Name two types of compound they found in the seawater that had not been there at the start.

_____ [2 marks]

(c) Explain why the experiment did not prove that this was how life was created.

_____ [2 marks]

B–A*

Carbon dioxide levels

1 The diagram below shows the carbon cycle.
Some of the processes have been replaced by letters.

(a) (i) Name process A. _____

(ii) Name process B. _____

(iii) Name process C. _____

(iv) Name process D. _____ [4 marks]

(b) Explain why process D appears twice. _____ [1 mark]

(c) Explain why burning biofuels is said to be 'carbon neutral'.

_____ [2 marks]

D–C

2 Biofuels and fossil fuels are both carbon-based fuels.

(a) What do we mean by a carbon-based fuel? _____ [1 mark]

(b) Explain why burning fossil fuels raises the levels of carbon dioxide in the atmosphere, but biofuels are considered to be neutral.

_____ [2 marks]

(c) Carbon dioxide is removed from the atmosphere by dissolving in seawater.
What problems might this process cause?

_____ [2 marks]

B–A*

Extended response question

Biofuels are fuels made from plant materials. In the UK in 2010, petrol and diesel had to contain at least 5% biofuel.

- Petrol and diesel made from crude oil are non-renewable energy sources.
- Biofuels are renewable sources of energy.
- Burning petrol, diesel and biofuels releases carbon dioxide into the atmosphere.
- Carbon dioxide is thought to cause global warming.
- Much of the carbon dioxide released into the atmosphere dissolves in the sea.

Use the information above, and your knowledge and understanding, to give the positive and negative environmental impacts of increasing the percentage of biofuels in petrol and diesel.

The quality of written communication will be assessed in your answer to this question.

_____

_____

_____

_____

_____

_____

_____

_____

_____

_____

_____

_____

_____

_____

_____

_____

_____

_____

_____

_____

_____

_____

_____

_____

_____

_____

_____

_____

_____ [6 marks]

Investigating atoms

1 (a) Complete this table (**i** to **iv**) about the mass and charge on subatomic particles.

Particle	Relative mass	Relative charge
proton	(i)	+1
(iii)	1	(ii)
electron	(iv)	−1

| 28 |
| **Si** |
| 14 |

[4 marks]

(b) Use the data in the box to help you answer these questions.

(i) How many protons does this element have? _____

(ii) How many neutrons does this element have? _____

(iii) Write the electronic structure of this element. _____ [3 marks]

2 Ernest Rutherford and his research team suggested, in 1909, that atoms were not solid but had a dense central nucleus. This is a diagram of the apparatus they used.

(a) What is produced at A? _____ [1 mark]

(b) What is B? _____ [1 mark]

(c) What is the purpose of the fluorescent screen?

_____ [1 mark]

(d) Explain how the deflections shown (by labels C to E) led Rutherford to suggest that atoms had a central nucleus.

_____

_____ [3 marks]

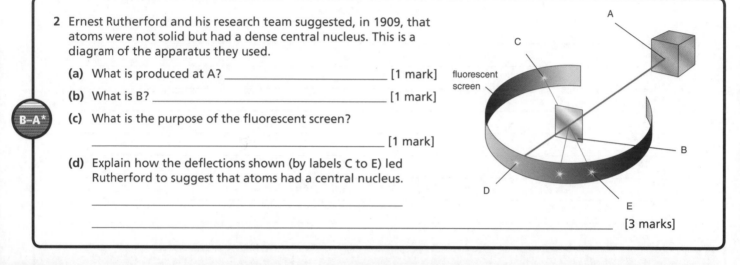

Mass number and isotopes

1 Many elements have different isotopes. Potassium has two isotopes: potassium-39 and potassium-40.

(a) What is an isotope?

_____ [1 mark]

(b) Describe how potassium-39 is different from potassium-40.

_____ [2 marks]

(c) In a sample of potassium atoms, 90 were found to be potassium-39, and 10 were potassium-40. Calculate the relative atomic mass of potassium.

_____

_____ [2 marks]

2 Scientists wanted to date some archaeological remains made of wood. Describe how they could use carbon-14 to find the age of the wood.

_____

_____ [2 marks]

Compounds and mixtures

1 Barium chloride solution is used to test for the presence of compounds containing sulfate ions. Some barium chloride was reacted with aluminium sulfate solution.

$$3BaCl_2(aq) + Al_2(SO_4)_3(aq) \rightarrow 3BaSO_4(s) + 2AlCl_3(aq)$$

(a) What does (aq) mean?

_____ [1 mark]

(b) Calculate the relative formula mass (M_r) of barium chloride.

(Relative atomic masses: Ba = 137, Cl = 35.5)

_____ [2 marks]

(c) What is the mass of one mole of barium chloride?

_____ [1 mark]

(d) Calculate the relative formula mass (M_r) of aluminium sulfate.

(Relative atomic masses: Al = 27, S = 32, O = 16)

_____ [2 marks]

D–C

Electronic structure

1 The diagram shows the electronic structure of a sodium atom.
Draw similar diagrams for the following atoms, writing the electronic structure under each diagram:

(a) calcium

2,8,1

[2 marks]

(b) helium

[2 marks]

D–C

(c) aluminium

[2 marks]

(d) fluorine

[2 marks]

(Atomic numbers: Al = 13, Ca = 20, F = 9, He = 2, Na = 11)

2 It is important to know how many outer electrons an element has.

(a) Use a copy of the periodic table to find out the number of outer electrons for the following elements:

(i) bromine (Br) _____ [1 mark]

(ii) strontium (Sr) _____ [1 mark]

(b) Explain why it is important to know the number of outer electrons of an element.

_____ [1 mark]

B–A*

(c) Noble gases (Group 0) are very unreactive elements. Explain why, in terms of their electronic structure.

_____ [2 marks]

Ionic bonding

1 The compound potassium fluoride (KF) is being investigated by some scientists. They have discovered the following facts about potassium fluoride.

property	
melting point	857 °C
boiling point	1502 °C
conduction as a solid	no
conduction as a solution	yes

(a) The scientists think potassium fluoride has ionic bonds. Give two reasons from the table for their conclusion.

_____

_____ [2 marks]

D–C

(b) A potassium atom has the electronic structure of 2,8,8,1. A fluorine atom has the electronic structure 2,7.

(i) Describe the electronic structure of a potassium ion.

_____ [1 mark]

(ii) Describe the electronic structure of a fluoride ion.

_____ [1 mark]

(c) Explain how the two ions form the compound potassium fluoride.

_____ [2 marks]

2 A student put some sodium chloride in a beaker and tested it to see if it would conduct electricity, and discovered it did not conduct. The student then added some water to the beaker and found that the solution now conducted electricity.

(a) Explain why solid sodium chloride does not conduct electricity but when dissolved in water it does.

_____ [2 marks]

B–A*

(b) The student took some molten sodium chloride and tested it for electrical conductivity. It was a conductor. Explain why molten sodium chloride can conduct electricity.

_____ [1 mark]

Alkali metals

1 Lithium, sodium and potassium are the first three elements of Group 1 or the alkali metals. They all have a single outer electron, and all react easily with water to form hydrogen gas and the metal hydroxide. Potassium has a violent reaction with water, and lithium just floats and fizzes in water.

(a) Write a word equation for the reaction of sodium with water. _____ [1 mark]

D–C

(b) Explain why all the Group 1 metals form positive ions with a 1+ charge.

_____ [2 marks]

(c) Caesium is another Group 1 metal. It is lower in the group than potassium. Suggest how it would react with water.

Explain your answer. _____ [2 marks]

2 Alkali metals are very common in the Earth's crust. Unlike iron and copper, which were first extracted thousands of years ago, potassium and sodium were only extracted two hundred years ago.

(a) Why was it possible for iron and copper to be extracted by man much earlier than sodium and potassium?

_____ [1 mark]

B–A*

(b) How are potassium and sodium extracted from their metal ores?

_____ [1 mark]

(c) Why was it only possible to extract them for the first time two hundred years ago?

_____ [1 mark]

Halogens

1 The diagram shows the electronic arrangement of a chlorine atom.

    (a) Complete the second diagram to show the electron arrangement of a chloride ion. [1 mark]

    (b) What is the charge on a chloride ion? _____ [1 mark]

    (c) Name the element with the same electronic structure as a chloride ion. _____ [1 mark]

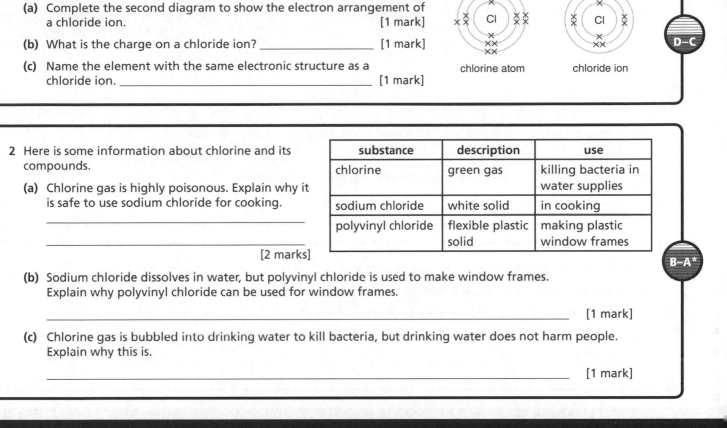

chlorine atom   chloride ion

D–C

2 Here is some information about chlorine and its compounds.

substance	description	use
chlorine	green gas	killing bacteria in water supplies
sodium chloride	white solid	in cooking
polyvinyl chloride	flexible plastic solid	making plastic window frames

    (a) Chlorine gas is highly poisonous. Explain why it is safe to use sodium chloride for cooking.

    _____

    _____

    [2 marks]

    (b) Sodium chloride dissolves in water, but polyvinyl chloride is used to make window frames. Explain why polyvinyl chloride can be used for window frames.

    _____ [1 mark]

    (c) Chlorine gas is bubbled into drinking water to kill bacteria, but drinking water does not harm people. Explain why this is.

    _____ [1 mark]

B–A*

Ionic lattices

1 The diagram shows part of a sodium chloride crystal lattice

    (a) How many chloride ions are attracted to each sodium ion? [1 mark]

    (b) Sodium chloride has a melting point of approximately 800 °C, but the molecule hydrogen chloride is a gas at room temperature. Explain why.

    _____ [2 marks]

    (c) Two electrodes, one positive and one negative, are placed in some molten sodium chloride, and a current passed through. Describe how the sodium and chloride ions react.

    _____

    _____ [3 marks]

● Na^+ ◯ Cl^-

D–C

2 Sodium sulfate is an ionic compound and has the formula Na_2SO_4.

    (a) Write the formulae of the two ions in sodium sulfate.

    _____ [2 marks]

    (b) Calculate the relative formula mass for sodium sulfate.

    (Relative atomic masses (A_r), Na = 23, O = 16, S = 32.)

    _____ [2 marks]

B–A*

Covalent bonding

1 Look at this diagram of two chlorine atoms.

(a) Draw a diagram to show how the two atoms will make a molecule of chlorine gas (Cl_2)

[2 marks]

(b) The atoms are joined together by a covalent bond. What is a covalent bond?

_____ [1 mark]

(c) Explain why non-metal elements form molecules with covalent bonds.

_____ [2 marks]

D–C

2 This diagram represents the structure of carbon dioxide.   O = C = O

(a) How many covalent bonds are made by each oxygen atom? _____ [1 mark]

(b) How many covalent bonds are made by the carbon atom? _____ [1 mark]

(c) The bond length of a C=O bond is 0.116 nm. How far apart are the two oxygen atoms' nuclei?

_____ [1 mark]

B–A*

Covalent molecules

1 The table shows some data about the halogens, Group 7.

(a) Room temperature is 20 °C. Which of the halogens are gases at room temperature? _____ [1 mark]

(b) How many outer electrons has a bromine atom? ___ [1 mark]

(c) Describe the trend in boiling points as the halogen molecule gets larger.

halogen	boiling point in °C	diameter of molecule in nm
fluorine	−220	0.28
chlorine	−34	0.39
bromine	58	0.46
iodine	114	0.53

_____ [1 mark]

(d) The boiling point of each halogen is affected by the intermolecular forces between the molecules. Describe, with reasons, how the intermolecular forces between halogen molecules change in the group.

_____

_____

_____ [3 marks]

D–C

2 Water molecules have a mass of 18. This is the same as fluorine molecules. Fluorine boils at −220 °C, but water has a boiling point of 100 °C.

(a) Explain why water boils at 100 °C instead of −220 °C.

_____ [2 marks]

(b) Suggest, with reasons, which halogen is likely to have intermolecular forces of about the same strength as water. (The table in question 1 may help.)

_____ [2 marks]

B–A*

Covalent lattices

1 The diagram shows the structures of diamond and graphite, two covalent lattices of carbon.

(a) How many covalent bonds does each carbon atom have in:

(i) diamond? _____ [1 mark]

(ii) graphite? _____ [1 mark]

(b) Diamond is one of the hardest known substances. Describe how its structure enables it to be very hard.

_____

_____ [2 marks]

(c) Graphite is often used as a lubricant. Describe how its structure enables it to be a good lubricant.

_____ [2 marks]

graphite   diamond

D–C

2 Diamond does not conduct electricity, but graphite does. This is because graphite has de-localised electrons.

(a) What are de-localised electrons?

_____ [1 mark]

(b) Explain how de-localised electrons allow graphite to conduct electricity.

_____

_____ [2 marks]

B–A*

Polymer chains

1 A bottle manufacturer has two different types of plastic bottle. One bottle is made from a thermosetting polymer, the other from a thermosoftening polymer. The thermosetting polymer is cheaper to produce.

(a) Describe the difference between thermosetting and thermosoftening polymers.

_____

_____ [2 marks]

(b) Suggest an environmental benefit from using the thermosoftening polymer bottle.

_____ [1 mark]

(c) Describe how the structure of the thermosetting polymer differs from the thermosoftening polymer.

_____

_____ [2 marks]

D–C

2 Polymer molecules are long chains of atoms held together by covalent bonds. There are weak attractions between the polymer chains. They have no definite melting point, instead they soften over a range of temperatures.

(a) Explain why they soften over a range of temperatures.

_____

_____ [2 marks]

(b) Suggest why polymers with shorter chains have lower melting ranges than polymers with longer chains.

_____ [1 mark]

B–A*

Metallic properties

1 The diagram on the right shows the metal lattice of a pure metal such as titanium.

When making a shape-memory alloy, titanium metal is mixed with nickel.

(a) Draw a diagram in the box on the far right to show the effect of adding nickel atoms to the titanium metal lattice. **[2 marks]**

(b) What is a shape-memory alloy?

_____ **[1 mark]**

(c) Describe how a shape-memory alloy is useful in making a dental brace.

_____ **[2 marks]**

D–C

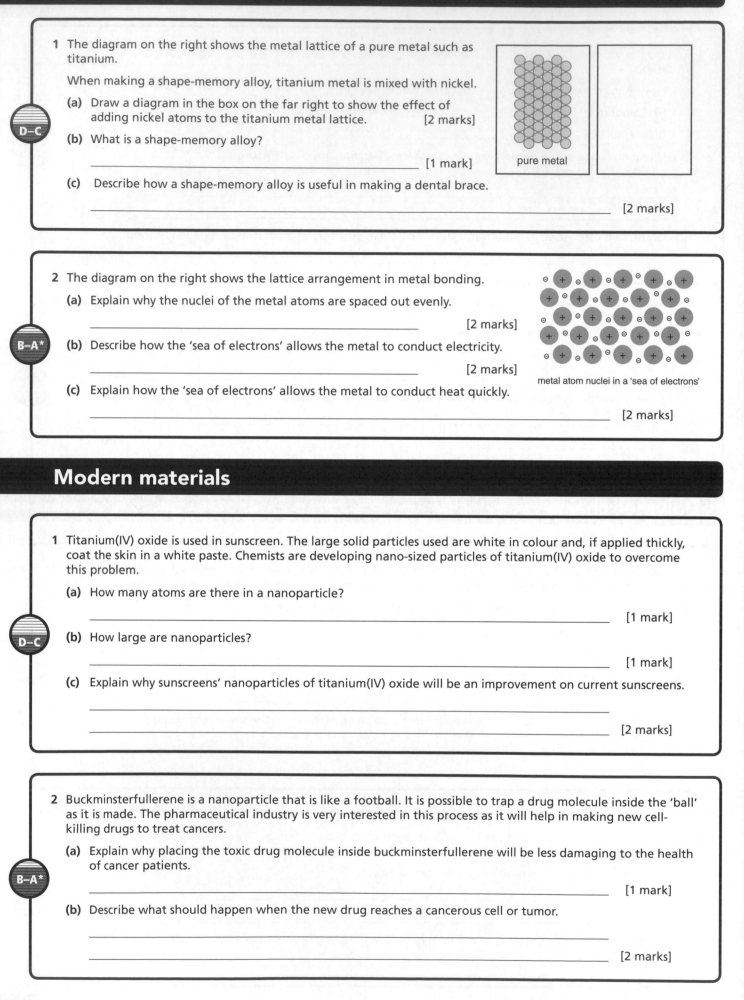

pure metal

2 The diagram on the right shows the lattice arrangement in metal bonding.

(a) Explain why the nuclei of the metal atoms are spaced out evenly.

_____ **[2 marks]**

(b) Describe how the 'sea of electrons' allows the metal to conduct electricity.

_____ **[2 marks]**

(c) Explain how the 'sea of electrons' allows the metal to conduct heat quickly.

_____ **[2 marks]**

B–A*

metal atom nuclei in a 'sea of electrons'

Modern materials

1 Titanium(IV) oxide is used in sunscreen. The large solid particles used are white in colour and, if applied thickly, coat the skin in a white paste. Chemists are developing nano-sized particles of titanium(IV) oxide to overcome this problem.

(a) How many atoms are there in a nanoparticle?

_____ **[1 mark]**

(b) How large are nanoparticles?

_____ **[1 mark]**

(c) Explain why sunscreens' nanoparticles of titanium(IV) oxide will be an improvement on current sunscreens.

_____

_____ **[2 marks]**

D–C

2 Buckminsterfullerene is a nanoparticle that is like a football. It is possible to trap a drug molecule inside the 'ball' as it is made. The pharmaceutical industry is very interested in this process as it will help in making new cell-killing drugs to treat cancers.

(a) Explain why placing the toxic drug molecule inside buckminsterfullerene will be less damaging to the health of cancer patients.

_____ **[1 mark]**

(b) Describe what should happen when the new drug reaches a cancerous cell or tumor.

_____

_____ **[2 marks]**

B–A*

Identifying food additives

1 Some crime scene investigators used chromatography to process some paint found at the scene of a serious hit-and-run accident. They wanted to know the manufacturer of the vehicle. Here is the chromatogram they produced.

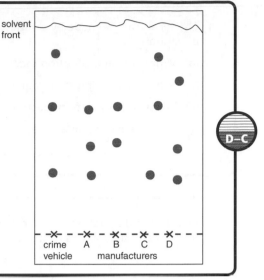

   (a) How many different chemicals does the sample from the hit-and-run car contain?

      _____ [1 mark]

   (b) All the spots are red. How many different red pigments are used by all the different manufacturers?

      _____ [1 mark]

   (c) Which manufacturer made the car?

      _____ [1 mark]

D–C

---

2 To help identify different pigments in a chromatogram the retention factor for a particular solvent and pigment is calculated.

   (a) Why is it important to calculate retention factors for different solvents for the same pigment?

      _____ [1 mark]

B–A*

   (b) A pigment travels 18 cm up a chromatogram, and the solvent has travelled up 27 cm. Calculate the retention factor for this pigment in this solvent.

      _____ [2 marks]

Instrumental methods

1 Gas Chromatography–Mass Spectrometry (GC-MS) is an instrumental method, used to determine the identity of different chemicals.

   (a) What is an instrumental method?

      _____ [1 mark]

   (b) Why are instrumental methods used?

      _____ [2 marks]

D–C

   (c) How does a mass spectrometer help analyse the results of the gas chromatogram?

      _____ [2 marks]

---

2 Mass spectrometers work by removing an electron from the molecule to be identified, and then finding the mass of the molecular ion. Sometimes in removing the electron the molecule is broken up into fragments.

   (a) What charge does a molecular ion always have?

      _____ [1 mark]

B–A*

   (b) What is the advantage of the molecule breaking into fragments when analysing two molecules with the same mass?

      _____ [2 marks]

Making chemicals

1 Chemicals can be made by a variety of different reactions such as neutralisation, oxidation, reduction, and precipitation.

   (a) What is a neutralisation reaction? _____ [1 mark]

   (b) What is a precipitation reaction? _____ [1 mark]

   (c) What is an oxidation reaction? _____ [1 mark]

   (d) Why are oxidation and reduction reactions sometimes described as redox reactions?

   _____ [2 marks]

D–C

2 In a chemical equation, the number of moles of each substance needed in the reaction is shown as a figure in front of the substance's formula in the balanced equation. Here is a balanced equation.

$$Na_2CO_3(s) + 2HCl(aq) \rightarrow 2NaCl + CO_2(g) + H_2O(l)$$

If one mole of sodium carbonate is used how many moles of:

   (a) hydrochloric acid are needed? _____ [1 mark]

   (b) sodium chloride are produced? _____ [1 mark]

   (c) carbon dioxide are produced? _____ [1 mark]

B–A*

Chemical composition

1 A student heated some copper oxide using this apparatus to reduce some copper oxide to copper metal.
The student weighed the 'boat', then the copper oxide and 'boat' before heating the copper oxide strongly. When cooled, the student then weighed the copper produced in the 'boat'.

burning hydrogen gas

hydrogen gas

porcelain 'boat' heat copper oxide

   (a) What was the mass of copper oxide used?

   _____ [1 mark]

   (b) What was the mass of copper produced?

   _____ [1 mark]

> Mass of boat = 15.44 g,
> boat and copper oxide = 19.42 g,
> boat and copper produced = 18.62 g

D–C

   (c) What was the mass of oxygen removed from the copper oxide?

   _____ [1 mark]

   (d) Use the masses to calculate the formula of copper oxide. Show your working.
   (Relative atomic masses, (A_r): Cu = 63.5, O = 16)

   _____ [3 marks]

2 A group of students wanted to find out if a hydrocarbon was an alkene. They burnt 3.36 g of the hydrocarbon and found it produced 10.56 g of carbon dioxide (CO_2), and 4.32 g of water (H_2O).
(Relative atomic masses, (A_r): C = 12, H = 1, O = 16)

   (a) What is the relative molecular mass of carbon dioxide?. _____ [1 mark]

   (b) What is the relative molecular mass of water? _____ [1 mark]

   (c) Calculate the empirical formula of the hydrocarbon. _____ [3 marks]

   (d) Is the hydrocarbon an alkene? Explain your answer.

   _____ [1 mark]

B–A*

Quantities

1 The reaction for the thermal decomposition of calcium carbonate to calcium oxide is shown by this equation.

$$CaCO_3(s) \rightarrow CaO(s) + CO_2(g)$$

(Relative atomic masses, (A_r): Ca = 40, C = 12, O = 16)

(a) Calculate the relative formula mass of calcium carbonate.

_____ [1 mark]

(b) Calculate the relative formula mass of calcium oxide.

_____ [1 mark]

(c) If 125 tonnes of calcium carbonate are used, what is the maximum theoretical yield of calcium oxide?

_____ [2 marks]

D–C

2 Lead iodide is obtained by precipitation of lead nitrate and sodium iodide solutions. The reaction shown by this equation:

$$Pb(NO_3)_2(aq) + 2NaI(aq) \rightarrow PbI_2(s) + 2NaNO_3(aq)$$

(Relative atomic masses, (A_r): Pb = 207, Na = 23, O = 16, I = 127, N = 14)

(a) Calculate the mass of sodium nitrate that can be made from 16.5 g of lead nitrate.

_____ [4 marks]

(b) Explain why this yield is unlikely.

_____ [1 mark]

B–A*

How much product?

1 A student planned to make some copper sulfate by reacting copper oxide with sulfuric acid.

$$CuO(s) + H_2SO_4(aq) \rightarrow CuSO_4(aq) + H_2O(\ell)$$

(Relative atomic masses, (A_r): Cu = 63.5, H = 1, S = 32, O = 16)

The student used 1.59 g of copper oxide with an excess of sulfuric acid. The student weighed the copper sulfate made, and found he had made 2.2 g of copper sulfate.

(a) What is meant by an excess? _____ [1 mark]

(b) Calculate the theoretical yield of copper sulfate. _____ [3 marks]

(c) Calculate the percentage yield obtained. _____ [1 mark]

D–C

2 A plant making sulfuric acid reacts sulfur with oxygen to make sulfur(VI) oxide according to this equation.

$$2S(s) + 3O_2(g) \rightarrow 2SO_3(g)$$

(Relative atomic masses, (A_r): S = 32, O = 16)

(a) Calculate the theoretical yield that can be obtained from burning 100 tonnes of sulfur.

_____ [3 marks]

B–A*

(b) Calculate the mass of oxygen required to burn the sulfur.

_____ [1 mark]

(c) If only 187.6 tonnes of sulfur(VI) oxide are made, what is the percentage yield?

_____ [2 marks]

Reactions that go both ways

1 Cobalt chloride paper can be used to test for the presence of water. It uses this reversible reaction:

$$CoCl_2 \cdot 6H_2O(s) \rightleftharpoons CoCl_2(s) + 6H_2O(\ell)$$

pink   blue

D–C

(a) What does this symbol $\rightleftharpoons$ mean?

_____ [1 mark]

(b) Describe the colour change when water is added to cobalt chloride paper.

_____ [1 mark]

(c) Explain why cobalt chloride paper may be dried and used again.

_____

_____ [2 marks]

Rates of reaction

1 A group of students wanted to measure the rate of reaction when calcium carbonate dissolves in hydrochloric acid.

$$CaCO_3(s) + HCl(aq) \rightarrow CaCl_2(aq) + H_2O(\ell) + CO_2(g)$$

They used the apparatus shown in the diagram. They placed 20 g of calcium carbonate in a conical flask then added 25 cm³ of dilute hydrochloric acid.

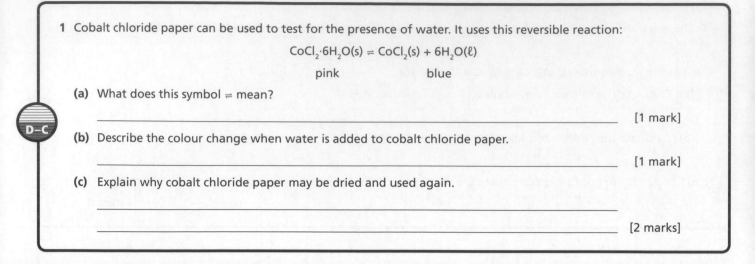

181.05 g

(a) Describe the measurements they need to take to follow the rate of the reaction.

_____ [2 marks]

Another group of students measured the volume of gas produced, and plotted a graph of their results.

D–C

(b) Explain why the graph is a curve.

_____

_____ [2 marks]

(c) What is the maximum volume of carbon dioxide obtained from the reaction?

_____ [1 mark]

2 (a) Use the graph above to calculate the rate of the reaction after 1.5 minutes.

_____

_____ [2 marks]

B–A*

(b) At the end of the experiment there is still calcium carbonate left in the flask. Explain why this has happened.

_____

_____ [2 marks]

Collision theory

1 Magnesium ribbon reacts with dilute hydrochloric acid like this:

$$Mg(s) + 2HCl(aq) \rightarrow MgCl_2 (aq) + H_2(g)$$

(a) If a 2 g piece of magnesium ribbon was reacted with an excess of hydrochloric acid, what would happen to the rate of reaction if the temperature was increased? Explain your answer.

_____ [2 marks]

(b) If a 2 g piece of magnesium ribbon was cut into six pieces then reacted with an excess of hydrochloric acid, what would happen to the rate of reaction? Explain your answer.

_____ [2 marks]

(c) If a 2 g piece of magnesium ribbon was reacted with a more dilute solution of hydrochloric acid, what would happen to the rate of reaction? Explain your answer.

_____ [2 marks]

D–C

2 Hydrogen gas can be made in the laboratory by reacting zinc metal with hydrochloric acid. Copper sulfate is often used as a catalyst.

(a) What is a catalyst? _____ [1 mark]

(b) A teacher uses a 2 cm³ cube of zinc metal for a first reaction, then repeats the reaction using the same-sized cube cut into eight smaller 1 cm³ cubes. Calculate the change in surface area, and so the change in the rate of the reaction. _____ [2 marks]

B–A*

(c) The teacher wants to check if the results of the second experiment are repeatable, so repeats the experiment using manganese oxide as the catalyst. Give two reasons why the results obtained cannot confirm the repeatability of the experiment. _____ [2 marks]

Adding energy

1 A student planned to make some iron sulfate by reacting 2 g of iron with an excess of sulfuric acid.

$$Fe(s) + H_2SO_4(aq) \rightarrow FeSO_4(aq) + H_2(g)$$

The student carried out the reaction at two different temperatures. The results are shown on the graph.

(a) What is the maximum volume of hydrogen gas the student can obtain? _____ [1 mark]

(b) The student decided to repeat the experiment at 40 °C.

   (i) Sketch on the graph what the results should look like.

        [2 marks]

   (ii) Explain your curve, using collision theory.

_____ [2 marks]

D–C

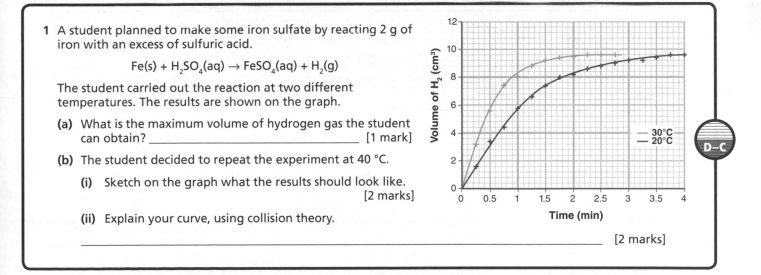

2 For a chemical reaction to take place, the reactant particles must have sufficient energy to react with each other when they collide. This energy is called the activation energy.

(a) Magnesium ribbon burns vigorously in air if briefly heated in a Bunsen flame. It does not burn without heating. Explain in terms of activation energy why only a little heat energy makes magnesium burn.

_____ [2 marks]

B–A*

(b) Suggest why increasing the temperature will increase the number of particles that have sufficient activation energy to react. _____ [2 marks]

Concentration

1 The graph shows the volume of gas produced when 3 g of aluminium reacted with hydrochloric acid at three different concentrations.

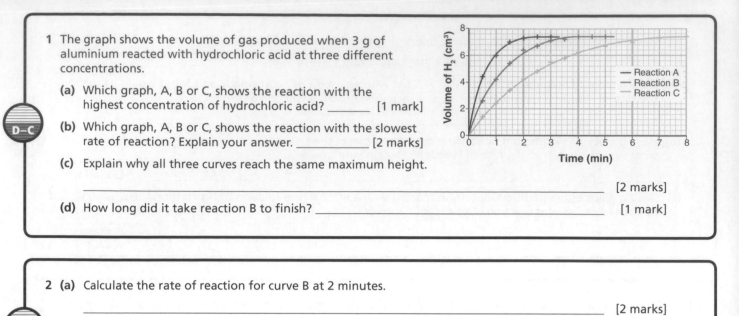

(a) Which graph, A, B or C, shows the reaction with the highest concentration of hydrochloric acid? _____ [1 mark]

(b) Which graph, A, B or C, shows the reaction with the slowest rate of reaction? Explain your answer. _____ [2 marks]

(c) Explain why all three curves reach the same maximum height.

_____ [2 marks]

(d) How long did it take reaction B to finish? _____ [1 mark]

---

2 (a) Calculate the rate of reaction for curve B at 2 minutes.

_____ [2 marks]

(b) If the concentration of hydrochloric acid in Reaction B was doubled, how long would it take the reaction to complete? Explain your answer.

_____ [2 marks]

Size matters

1 A student looked at burning iron in a Bunsen flame. The student tried burning a small nail, some iron wool, and some iron filings in a Bunsen flame. Here are the observations of the experiment.

type of iron	observations
nail	glowed red, did not burn, did not change size
iron wool held by tongs	bright red glow that spread through the wool, as it spread the wool disappeared, brown bits appeared on the heat mat
iron filings sprinkled into flame	bright sparks as filings sprinkled onto flame, brown bits on heat mat

(a) Which of the three types of iron reacted fastest? _____ [1 mark]

(b) Explain your answer using collision theory.

_____ [2 marks]

(c) Which of the types of iron did not react? What evidence is in the table for this?

_____ [2 marks]

---

2 Ships use fuel oil to heat their boilers, at normal speeds they spray fuel oil into the boilers. To rapidly increase their speed, they use additional sprayers that produce a much finer spray.

(a) Suggest why ships use sprayed fuel oil rather than liquid fuel oil.

_____ [2 marks]

(b) Explain why ships use a finer spray when they want to heat the boilers up faster and increase speed.

_____ [2 marks]

Clever catalysis

1 Hydrogen peroxide decomposes slowly to form water and oxygen. It can be speeded up by using manganese(IV) oxide as a catalyst.

A student wanted to prove that manganese(IV) oxide was a catalyst, so did an experiment. Here are the results.

volume of hydrogen peroxide in cm³	mass of manganese(IV) oxide at the start in grams	mass of manganese(IV) oxide at the end in grams	time for 50 cm³ of oxygen gas to be produced in seconds.
100	0.0	0.0	250
100	2.3	2.3	50

(a) By how much did the manganese(IV) oxide affect the rate of reaction? Explain your answer.

_____ [2 marks]

(b) What evidence is there that the manganese(IV) oxide is a catalyst? Explain your answer.

_____ [2 marks]

(c) Suggest how the manganese(IV) oxide may have affected the reaction rate.

_____ [2 marks]

D–C

2 Many motor vehicles now have catalytic converters fitted. Vehicles fitted with them cannot use petrol that contains lead.

(a) Explain why a car fitted with a catalytic converter cannot use leaded petrol.

_____ [2 marks]

(b) Catalytic converters are expected to last for at least 100 000 kilometres of use with unleaded petrol. Suggest why after this a catalytic converter may no longer work.

_____ [2 marks]

B–A*

Controlling important reactions

1 Ammonia is a very useful gas. It is used to make fertilisers. Controlling the rate of the reaction is very important to maximise the ammonia made, whilst keeping the costs economic. Ammonia is made by a reversible reaction between hydrogen and nitrogen.

(a) Write a word equation for the manufacture of ammonia.

_____ [1 mark]

(b) The reaction uses an iron catalyst. What does this suggest about the rate of the reaction between nitrogen and hydrogen? _____ [1 mark]

(c) At 450 °C only 20% of the nitrogen and hydrogen become ammonia. Explain why a temperature of 450 °C is used for the reaction.

_____ [2 marks]

(d) Explain why a pressure of 200 atmospheres is used for the reaction.

_____ [2 marks]

D–C

2 In industry ammonia is made at 450 °C and at a pressure of 200 atmospheres. The speed of reaction is more important than the yield.

(a) Use your knowledge of the Haber Process to suggest why yield is not an important economic consideration.

_____ [2 marks]

(b) Explain why this combination of temperature and pressure is chosen so that the process is economic.

_____ [3 marks]

B–A*

The ins and outs of energy

1 A student dissolved 5 g of ammonium nitrate in 50 cm³ of water. The temperature of the solution fell by 6 °C.

   (a) What type of reaction is this? _____ [1 mark]

   (b) Explain why the temperature dropped.

   _____ [2 marks]

   (c) The diagram shows a self-heating can of coffee. Which two chemicals must react for the coffee to start to heat up?

   _____ [1 mark]

   (d) Where does the heat energy come from to heat the coffee?

   _____ [2 marks]

D–C

coffee product

insert

quicklime

foil separator

water

plastic button

---

2 The reaction of anhydrous copper sulfate with water is a reversible reaction.

anhydrous copper sulfate + water ⇌ hydrated copper sulfate

   (a) What do we mean by a reversible reaction?

   _____ [1 mark]

   (b) What is meant by the forward reaction?

   _____ [1 mark]

   (c) If the forward reaction is exothermic, explain why the backward reaction is endothermic.

   _____ [2 marks]

B–A*

Acid-base chemistry

1 Alkalis and bases can both neutralise an acid.

   (a) Describe the difference between an alkali and a base.

   _____ [1 mark]

   (b) When sodium hydroxide solution reacts with nitric acid it forms two new compounds. Complete the equation to show the two products.

   sodium hydroxide + nitric acid → _____ + _____ [1 mark]

   (c) Copper carbonate can act as a base and react with hydrochloric acid. Complete the equation to show the products formed.

   copper carbonate + hydrochloric acid → _____ + _____ + _____ [2 marks]

D–C

---

2 A teacher demonstrated the dissolving of ammonia gas (NH_3) in water to some students. The water had some universal indicator in it. The water turned from green at the start to blue at the end.

   (a) What did the experiment show about ammonia gas?

   _____ [1 mark]

   (b) The ammonia dissolved in the water to form a positive and a negative ion. Write the formulae of the two ions formed.

   _____ [2 marks]

   (c) Explain why the solution made is able to neutralise acids.

   _____ [1 mark]

B–A*

C2 Chemistry

Making soluble salts

1 A student wanted to make some cobalt sulfate. The student planned to use cobalt metal, but a friend said this would be too hazardous and suggested using cobalt oxide instead.

   (a) Name the acid needed to make cobalt sulfate. _____ [1 mark]

   (b) The student added some cobalt oxide to the acid.

      (i) What should the student do to help the cobalt oxide and acid to react?

      _____ [1 mark]

      (ii) Suggest two ways that the student could use to show all the acid had been used up.

      _____ [2 marks]

   (c) How could the student make sure that the cobalt sulfate made was pure, and had no cobalt oxide left in it?

   _____ [1 mark]

   (d) How could the student obtain a solid sample of cobalt sulfate crystals?

   _____ [2 marks]

(D–C)

2 Calcium sulfate is used to make Plaster of Paris. A teacher decided to make some Plaster of Paris by reacting some pieces of calcium carbonate with sulfuric acid.

   (a) What did the teacher expect to see when the calcium carbonate reacted with the sulfuric acid?

   _____ [1 mark]

   (b) This happened for a short while and then stopped. Explain why the reaction stopped.

   _____ [2 marks]

(B–A*)

Insoluble salts

1 Calcium carbonate obtained from limestone or chalk is too impure to be used in kitchen cream cleaning liquids. Instead it is obtained by using this reaction.

$$CaCl_2(aq) + Na_2CO_3(aq) \rightarrow CaCO_3(s) + 2NaCl(aq)$$

   (a) Name the products in the reaction. _____ [1 mark]

   (b) After the reaction is complete, how could you separate the calcium carbonate from the solution?

   _____ [1 mark]

   (c) Explain how you could ensure that the calcium carbonate would be pure when dried.

   _____ [1 mark]

(D–C)

2 Silver nitrate ($AgNO_3$) solution is used to test solutions thought to contain halogen compounds or halides such as chlorides, bromides and iodides.

   (a) If a solution of calcium iodide (CaI_2) is tested with silver nitrate solution:

      (i) What would you see happen? _____ [1 mark]

      (ii) What precipitate would be formed? _____ [1 mark]

(B–A*)

   (b) If a solution containing magnesium chloride ($MgCl_2$) is tested with silver nitrate solution, what colour would you expect the precipitate to be? _____ [1 mark]

   (c) Write a balanced symbol equation for the reaction of silver nitrate solution with magnesium chloride solution.

   _____ [2 marks]

Ionic liquids

1 Potassium dichromate ($K_2Cr_2O_7$) is a yellow ionic solid that dissolves in water to form potassium ions (K^+) and dichromate ions ($Cr_2O_7^{2-}$). The yellow colour is caused by the dichromate ions. If an electric current is passed through a filter paper soaked in water with a pipette of potassium dichromate added as shown on the right, this is what happens.

(a) What is the name given to the process of splitting a compound using electricity?

_____ [1 mark]

(b) Suggest why the yellow colour travels towards the positive electrode or anode.

_____ [2 marks]

(c) Explain what will happen to a potassium ion when it reaches the negative electrode.

_____ [2 marks]

2 (a) Explain why the movement of the dichromate and potassium ions allows electricity to flow round the circuit.

_____ [2 marks]

(b) A student tried to do this experiment but connected the electrodes to a power pack that only provided alternating current. The circuit conducted electricity, but the colour stayed in the middle of the filter paper. Explain why?

_____ [2 marks]

Electrolysis

1 The diagram shows how copper is purified using electrolysis.

(a) Which electrode has the impure copper on? _____ [1 mark]

(b) What happens to the impure copper at this electrode?

_____ [2 marks]

(c) Explain why the electrolyte solution has to contain copper ions.

_____ [1 mark]

(d) What are substances X? _____ [1 mark]

2 When molten sodium chloride is electrolysed, the sodium ions [$Na^+(\ell)$] are attracted to the cathode where they gain electrons. This can be represented by the half-equation:

$$Na^+(\ell) + e^- \rightarrow Na(\ell)$$

(a) Why is this called a half-equation?

_____ [1 mark]

(b) Write a similar half-equation for the reaction happening at the anode with chloride ions.

_____ [1 mark]

(c) Explain why the reaction at the anode is called an oxidation reaction?

_____ [1 mark]

(d) Write a balanced equation showing the overall reaction.

_____ [1 mark]

Extended response question

Sodium hydroxide is made by the electrolysis of brine (sodium chloride) at room temperature. Sodium chloride is an ionic compound containing sodium ions (Na^+) and chloride ions (Cl^-).

Sodium hydroxide can also be made by the electrolysis of molten sodium chloride and calcium chloride at 600 °C, and then reacting the sodium produced with water.

The figure shows the apparatus used to electrolyse the brine.

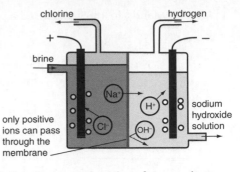

Use information in the box, and your knowledge and understanding of this process, to answer this question.

Explain, as fully as you can, why sodium hydroxide is made by the electrolysis of brine rather than from molten sodium chloride.

The quality of written communication will be assessed in your answer to this question.

[6 marks]

Ordering elements

1 John Newlands, in 1865, proposed a periodic table of elements. The first section is printed below.

H	Li	Be	B	C	N	O
F	Na	Mg	Al	Si	P	S
Cl	K	Ca	Cr	Ti	Mn	Fe
Co/Ni	Cu	Zn	Y	In	As	Se

D–C

(a) Give two differences between the modern periodic table and Newland's.

_____ [2 marks]

(b) Suggest two reasons why the table was not accepted by other scientists.

_____ [2 marks]

(c) In 1869, Mendeleev proposed a table that was accepted. He used the same method as Newlands but made two changes. What were the changes?

_____ [2 marks]

2 In 1869, Mendeleev predicted the existence of the element eka-silicon. He predicted that it had an atomic mass of 72, and would be coloured grey. The element was discovered in 1875.

(a) What is the name used today for eka-silicon? _____ [1 mark]

(b) Use the periodic table to help you with this question. How good was the prediction of its atomic mass?

_____ [1 mark]

B–A*

(c) Suggest how its discovery helped persuade scientists that Mendeleev's periodic table would work.

_____ [1 mark]

(d) The elements in the table are organised in mass number order.
Explain why tellurium and iodine do not follow this rule.

_____ [1 mark]

(e) There is one group missing from the original of Mendeleev's periodic table. Which group is it, and why was it omitted?

_____ [2 marks]

The modern periodic table

1 This is a diagram of the electronic structure of sodium.

(a) Which group of the periodic table is sodium in? Explain your answer.

_____ [2 marks]

(b) Draw a similar diagram for argon.

D–C

2,8,1

_____ [1 mark]

(c) Explain why sodium is a highly reactive element, and argon is a very unreactive element.

_____ [2 marks]

(d) Fluorine and chlorine both react similarly with sodium to form sodium fluoride (NaF) and sodium chloride (NaCl). Explain why fluorine and chlorine react similarly with sodium.

_____ [1 mark]

Group 1

1 Group 1 metals, lithium, sodium and potassium, all react easily with water.

(a) Which of the three metals is most reactive? _____ [1 mark]

(b) Explain why this metal is most reactive. Your answer should refer to the differences in electronic structure of the elements. _____ [3 marks]

(c) Describe what you would see if some sodium was added to a trough of water.

_____ [3 marks]

(d) The reaction can be represented as a chemical equation: $2Na(s) + 2H_2O(\ell) \rightarrow 2NaOH(aq) + H_2(g)$

(i) What is the name of NaOH? _____ [1 mark]

(ii) Calculate the relative formula mass of NaOH. [A_r: Na = 23, H = 1, O = 16]

_____ [1 mark]

D–C

2 Rubidium and caesium are also in Group 1. Explain why they are not used in schools to show the trend in reactivity. _____ [2 marks]

B–A*

Transition metals

1 Transition metal compounds are used to colour glass used to make ornamental glassware.

(a) Why are transition metal compounds used to colour glass? _____ [1 mark]

(b) Iron can react with sulfuric acid to make two different compounds, green iron(II) sulfate and brown iron(III) sulfate. Explain why there are two types of iron sulfate. _____ [2 marks]

(c) Platinum is used as a catalyst in car emission systems.

(i) What is a catalyst? _____ [1 mark]

(ii) Why is it used in car emission systems? _____ [2 marks]

D–C

2 The table below shows the cost in 2011 of some transition metals, their abundance on Earth, and the chemical reactions that can be catalysed by the metal. The table is arranged in order of increasing cost.

Metal	Cost in $/kg	Used as a catalyst for	Abundance in Earth's crust (%)
iron	1	making ammonia	5.63
nickel	19	hydrogenation of vegetable oils	0.0084
ruthenium	5 600	making ammonia	1×10^{-7}
osmium	12 000	making ammonia	1.5×10^{-7}
rhodium	57 000	car emission systems	1×10^{-7}
platinum	60 000	car emission systems	5×10^{-7}

(a) What is the relationship between cost and abundance? _____ [1 mark]

(b) Which pair of elements do not fit the pattern? _____ [1 mark]

(c) Suggest why rhodium, which is more plentiful than osmium, costs more?

_____ [2 marks]

(d) Ruthenium and osmium were both used to catalyse the formation of ammonia. Suggest why iron is used, despite being poorer as a catalyst. _____ [2 marks]

(e) Explain why recycling a car emission system is both economically and environmentally good.

_____ [2 marks]

B–A*

Group 7

1 The Group 7 elements, fluorine, chlorine and bromine, react with sodium. Fluorine reacts very vigorously, chlorine reacts vigorously, and then bromine reacts slowly. [Atomic numbers F = 9, Cl = 17, Br = 35]

D–C

(a) How many outer electrons does a halogen atom have? _____ [1 mark]

(b) What is the charge on a chlorine ion? _____ [1 mark]

(c) Use electronic structures to explain why fluorine is the most reactive halogen element.

_____

[3 marks]

2 Hydrogen chloride is a sharp-smelling gas made when salt reacts with an acid such as vinegar (ethanoic acid). The hydrogen and chlorine atoms are joined together by a covalent bond.

(a) Draw a dot and cross diagram to show the bonding in a hydrogen chloride molecule. [Atomic numbers H = 1, Cl = 17]

[2 marks]

B–A*

(b) When hydrogen chloride dissolves in water, it produces an acidic solution. Explain what happens to the molecule when it dissolves in water that makes the solution acidic.

_____

[2 marks]

(c) Explain why hydrogen chloride gas does not conduct electricity, but hydrogen chloride solution can.

_____

[2 marks]

Hard and soft water

1 A student investigated some water samples by finding out how much soap solution was needed to form a permanent lather, before boiling and after boiling the water. The results are shown in the table.

D–C

(a) Which samples are soft water?
_____ [1 mark]

(b) Which sample is temporary hard water? Explain your answer.
_____ [2 marks]

(c) Which sample contains both permanent and temporary hard water? Explain your answer.

water sample	volume of soap solution in cm³ needed to form a permanent lather		observations
	before boiling	after boiling	
A	2	2	none
B	5	5	scum produced before and after boiling
C	10	2	scum before boiling, no scum after boiling
D	10	7	scum produced before and after boiling
E	3	3	none

_____

[2 marks]

(d) How can both permanent and temporary hard water be softened. Give two methods.

_____

[2 marks]

2 Scum is a white insoluble solid formed when soap reacts with calcium ions in water. Boiling temporary hard water prevents scum forming. This equation shows why. $Ca(HCO_3)_2(aq) \rightarrow CaCO_3(s) + H_2O + CO_2(g)$

B–A*

(a) What is the missing state symbol? _____ [1 mark]

(b) Name the compound $Ca(HCO_3)_2$. _____ [1 mark]

(c) Explain why $CaCO_3$ formation stops the formation of scum when using soap.

_____

[2 marks]

Safe drinking water

1 A water filter jug has three parts: a carbon filter, an ion exchange resin, and a silver mesh.

(a) Explain the purpose of the carbon filter.

_____ [1 mark]

(b) Explain how ion exchange resins remove metal ions from water.

_____ [2 marks]

(c) Why does the water pass through a silver mesh screen?

_____ [1 mark]

D–C

2 Desalination of sea water can be undertaken in the laboratory using this apparatus.

(a) Explain what is meant by desalination.

_____ [1 mark]

(b) Describe how the seawater is desalinated by this process.

_____

_____ [4 marks]

thermometer
water out
condenser
water in
seawater
heat
beaker
distilled water

B–A*

(c) Explain why hot, dry countries surrounded by sea do not use desalination plants to make drinking water.

_____ [1 mark]

Energy from reactions

1 The diagram shows how a student set up an experiment to find the energy released when ethanol burns.

The student burnt 2 g of ethanol, and the water temperature rose by 7 °C.

(a) Calculate the energy gained by the water during the experiment. [Assume that 1 cm³ of water has a mass of 1 g, and that the specific heat capacity of water is 4.2 J/g °C]

_____ [2 marks]

(b) A friend suggested that there was a problem with the measurement of the temperature. Suggest what the problem is, and how it could be corrected.

_____ [2 marks]

(c) When the student compared the result to one on the internet, the internet value was much larger. Explain why the internet value was much larger.

_____ [2 marks]

200 cm³ of water

ethanol

D–C

2 (a) Calculate the relative formula mass of ethanol (C_2H_5OH). [A_r C = 12, O = 16, H = 1]

_____ [1 mark]

(b) Use your answers to questions 1 (a) and 2 (a) to calculate what the energy released would be if 1 mole of ethanol had been burnt.

_____

_____ [3 marks]

B–A*

Energy from bonds

1 Ethane burns in oxygen to make carbon dioxide and water vapour. This equation shows the reaction.

$$2C_2H_6(g) + 7O_2(g) \rightarrow 4CO_2(g) + 6H_2O(\ell)$$

(a) The reaction is exothermic.
What does this mean?

_____

[1 mark]

Bond energies

bond	H–H	C–H	O=O	O–H	C=O	C–C
bond energy (kJ/mol)	436	413	496	463	743	346

(b) Explain in terms of bonds, why the reaction is exothermic.

_____ [2 marks]

(c) Calculate the energy change when 1 mole of ethane burns in oxygen.

_____ [3 marks]

D–C

2 A student researched the energy released when bonds were broken. Breaking the first C–H bond in a molecule of methane (CH_4) required 435 kJ mol^{-1}, but breaking all four C–H bonds in methane required 1662 kJ mol^{-1}.

(a) Calculate the average bond energy of a C–H bond in methane.

_____ [1 mark]

(b) Explain why this value is different to the energy needed to break the first C–H bond in methane.

_____ [2 marks]

B–A*

Energy saving chemistry

1 Hydrogen is a 'wonder fuel' for cars. When burnt, it produces only water vapour so is non polluting.

$$H_2(g) + O_2(g) \rightarrow H_2O(\ell)$$

(a) Balance the chemical equation for the reaction of hydrogen with oxygen. [1 mark]

(b) Describe how hydrogen can be obtained for use as a fuel.

_____ [2 marks]

(c) Although burning hydrogen is pollution free, explain why using hydrogen will still cause acid rain.

_____ [2 marks]

(d) Suggest a problem that might occur in the event of a car crash where cars use hydrogen as a fuel.

_____ [1 mark]

D–C

2 Hydrogen can be used in a fuel cell to produce electricity. The catalyst used is platinum.

(a) Describe the function of the catalyst with the hydrogen gas.

_____ [1 mark]

(b) Describe what happens at the cathode. _____ [1 mark]

(c) Suggest two reasons why fuel cells are not in common use.

_____

_____ [2 marks]

B–A*

Analysis – metal ions

1 One way of analysing the metal ion present in a solution is to add sodium hydroxide solution to the solution.

(a) A student did this and obtained some different-coloured precipitates.
Suggest the metal present in a solution that gives a:

(i) blue precipitate _____ [1 mark]

(ii) green precipitate _____ [1 mark]

(iii) brown precipitate _____ [1 mark]

(b) The student noticed that some solutions did not give a precipitate. To which Group did these metals belong?

_____ [1 mark]

(c) One solution produced a white precipitate, but this disappeared when more sodium hydroxide was added. Which metal ion was this? _____ [1 mark]

D–C

Analysis – non-metal ions

1 A laboratory technician found some unlabelled bottles of white solids on a shelf. The bottles probably contained sodium chloride, sodium iodide, sodium bromide, sodium sulfate and sodium carbonate. To find out which bottle contained which chemical, the technician dissolved a small amount of each solid in some water and then added a few drops of nitric acid and then silver nitrate solution. The results are shown in the table.

solid	observation
A	yellow precipitate
B	white precipitate
C	no reaction
D	produced a gas
E	cream precipitate

(a) Which solid is:

(i) sodium chloride? _____ [1 mark]

(ii) sodium bromide? _____ [1 mark]

(iii) sodium carbonate? _____ [1 mark]

(b) Describe a test that would show that C was a sulfate, and describe what you would see.

_____

_____ [2 marks]

D–C

2 Barium sulfate is a highly toxic compound. It is given to thousands of patients in hospital as a drink every day before having an X-ray.

(a) Explain why it is safe for patients to drink despite being toxic.

_____

_____ [2 marks]

(b) Explain why barium sulfate is given before an X-ray.

_____

_____ [2 marks]

B–A*

Measuring in moles

1 How many moles are there in:

   (a) 129 g of $CoCl_2$? _____ [2 marks]

   (b) 12.6 g of $FeCl_2$? _____ [2 marks]

   (c) 48 g of CH_4? _____ [2 marks]

   [A_r Co = 59, Fe = 56, Cl = 35, C = 12, H = 1]

2 What is the concentration in moles per decimetre (mol/dm³) of these solutions?

   (a) 65 g of $CoCl_2$ in 500 cm³ of water _____ [2 marks]

   (b) 25.2 g of $FeCl_2$ in 200 cm³ of water _____ [2 marks]

   (c) 7.2 g of HCl in 500 cm³ of water _____ [2 marks]

   [A_r Co = 59, Fe = 56, Cl = 35, C = 12, H = 1]

Analysis – acids and alkalis

1 A student carried out a titration by reacting 25 cm³ of sodium hydroxide solution with some 0.1 mol/dm³ hydrochloric acid to find the molarity of the alkali.

   Here are the results.

Titration	rough	1	2	3	mean
volume of acid used in cm³	20.5	20.2	20.3	20.2	

   (a) Calculate the mean of the readings. [1 mark]

   (b) Calculate the number of moles of hydrochloric acid used.

   _____ [2 marks]

   (c) If the equation for the reaction is

   $$HCl(aq) + NaOH(aq) \rightarrow NaCl(aq) + H_2O(\ell)$$

   What is the number of moles in the 25 cm³ of sodium hydroxide? _____ [1 mark]

   (d) Calculate the concentration of the sodium hydroxide solution.

   _____ [2 marks]

2 Sulfuric acid can be used to titrate an alkali such as ammonia solution. The reaction can be represented as:

   $$2NH_3(aq) + H_2SO_4(aq) \rightarrow (NH_4)_2SO_4(aq)$$

   If 24.5 cm³ of 1 mol/dm³ of sulfuric acid was needed to neutralise the ammonia solution, work out:

   (a) the number of moles of sulfuric acid used. _____ [2 marks]

   (b) the number of moles of ammonia present. _____ [2 marks]

   (c) the mass of ammonium sulfate in the solution at the end.

   _____ [2 marks]

   [A_r N = 14, O = 16, S = 32, H = 1]

Dynamic equilibrium

1 This equation represents the equilibrium between sulfur dioxide, sulfur trioxide and oxygen. The reaction is exothermic.

$$2SO_2(g) + O_2(g) \rightarrow 2SO_3(g)$$

(a) If the temperature is increased, what will happen to the mass of oxygen? Explain why.

_____ [2 marks]

(b) If a catalyst is used what will happen to the mass of sulfur trioxide (SO_3)? Explain why.

_____ [2 marks]

D–C

2 If the pressure of the equilibrium is increased, what will happen to the mass of the reactants? Explain why.

_____ [2 marks]

B–A*

Making ammonia

1 This graph shows the effect of pressure and temperature on the conversion of nitrogen and hydrogen to ammonia, the Haber process.

(a) At what temperature is the conversion rate the highest?

_____ [1 mark]

(b) What happens to the conversion rate when the pressure is increased?

_____ [1 mark]

(c) Use the graph to find the conversion rate when the temperature is 400 °C, with a pressure of 500 atmospheres.

_____ [1 mark]

D–C

Graph: Conversion to NH₃ per cent (y-axis, 0–100) vs Pressure/atmospheres (x-axis, 0–600). Curves labelled 200 °C, 400 °C, 500 °C.

2 If the chosen temperature and pressure for the reaction are 450 °C and 200 atmospheres:

(a) estimate the conversion rate for these conditions

_____ [1 mark]

(b) use the graphs and your knowledge to explain why these conditions are used.

_____

_____

_____

_____ [4 marks]

B–A*

Alcohols

1 Here is a diagram of a molecule.

$$HO-\overset{\overset{\displaystyle H}{|}}{\underset{\underset{\displaystyle H}{|}}{C}}-\overset{\overset{\displaystyle H}{|}}{\underset{\underset{\displaystyle H}{|}}{C}}-\overset{\overset{\displaystyle H}{|}}{\underset{\underset{\displaystyle H}{|}}{C}}-\overset{\overset{\displaystyle H}{|}}{\underset{\underset{\displaystyle H}{|}}{C}}-H$$

(a) What is the general name given to this type of molecule?

_____ [1 mark]

(b) This molecule burns in the air releasing lots of energy. Name the two compounds formed when this molecule burns.

_____ [2 marks]

D–C

---

2 (a) One of these molecules is called ethanol. It has the formula C_2H_5OH. It reacts with oxygen to form a carboxylic acid.

Complete the diagram to show the carboxylic acid formed.

$$H-\overset{\overset{\displaystyle H}{|}}{\underset{\underset{\displaystyle H}{|}}{C}}-\overset{\overset{\displaystyle H}{|}}{\underset{\underset{\displaystyle H}{|}}{C}}-OH \; + \; O_2 \; \longrightarrow \qquad\qquad + H_2O$$

[2 marks]

(b) Name the carboxylic acid you have drawn in (a).

_____ [1 mark]

B–A*

Carboxylic acids

1 When a carboxylic acid reacts with an alcohol there is a condensation reaction like this.

$$H-\overset{\overset{\displaystyle H}{|}}{\underset{\underset{\displaystyle O}{\|}}{C}}-C-OH \; + \; HO-\overset{\overset{\displaystyle H}{|}}{\underset{\underset{\displaystyle H}{|}}{C}}-\overset{\overset{\displaystyle H}{|}}{\underset{\underset{\displaystyle H}{|}}{C}}-H \; \longrightarrow \; H-\overset{\overset{\displaystyle H}{|}}{\underset{\underset{\displaystyle O}{\|}}{C}}-C-O-\overset{\overset{\displaystyle H}{|}}{\underset{\underset{\displaystyle H}{|}}{C}}-\overset{\overset{\displaystyle H}{|}}{\underset{\underset{\displaystyle H}{|}}{C}}-H \; + \; H_2O$$

ethanoic acid · · · · · ethanol · · · · · $\longrightarrow$ · · · · · ethyl ethanoate · · · + · · · water

The reaction is exothermic, and the water made is often produced as steam.

(a) Suggest why the reaction is called a condensation reaction.

_____ [1 mark]

(b) What type of substance is ethyl ethanoate? _____ [1 mark]

(c) Explain why ethyl ethanoate is often added to foods and perfumes.

_____ [2 marks]

D–C

---

2 Carboxylic acids like ethanoic acid have similar reactions to hydrochloric acid and sulfuric acid. Here is some information about these three acids.

name of acid	pH when dissolved in water	name of salt that the acid forms
hydrochloric	1	chloride
sulfuric	1	sulfate
ethanoic	3	ethanoate

(a) Name the products that would be made if ethanoic acid was reacted with sodium carbonate.

_____ [3 marks]

(b) When dissolved in water, ethanoic acid only has a pH of 3. What do we call acids that have a pH between 3 and 6? _____ [1 mark]

(c) Explain why the pH of ethanoic acid is different from that of hydrochloric acid.

_____ [2 marks]

B–A*

Extended response question

In this question you will be assessed on using good English, organising information clearly and using specialist terms where appropriate.

The student wanted to find the energy in joules that could be released by burning 1 mole or 46 g of ethanol. The student had only 2 g of ethanol to burn.

Describe how the student could do this by heating water.

In your description you should include:

- the names of pieces of apparatus used

- the names of the substances used

- a risk assessment.

The quality of written communication will be assessed in your answer to this question.

_____

_____

_____

_____

_____

_____

_____

_____

_____

_____

_____

_____

_____

_____

_____

_____

_____

_____

_____

_____

_____

_____

_____

_____

_____ [6 marks]

1   I can name and position the three subatomic particles in an atom of any of the first 20 elements using the periodic table, including writing the electronic structures ☐

2   I can describe how non-metal compounds share electrons to form molecules, and how the atoms of metal and non-metal compounds transfer electrons to form ions ☐

3   I can understand and use a symbol equation to determine the number of atoms in a reaction, and calculate the mass of a reactant or product from the masses of the other reactants and products ☐

4   I can explain how limestone can be used to produce quicklime, limewater, cement, mortar and concrete ☐

5   I can evaluate the social, economic and environmental impacts of quarrying limestone and metal ores ☐

6   I can describe how metals – such as iron, copper and aluminium – are extracted from their ores by reduction ☐

7   I can use the reactivity series to identify whether electrolysis or heating with carbon is the best method of extraction for a metal ☐

8   I can explain why electrolysis is an expensive method of extraction, and list the benefits of recycling metals ☐

9   I can describe how copper can be obtained from low-grade ores, by phytomining or bioleaching, and explain the environmental and economic benefits of these processes ☐

10 I can describe the differences between iron, low- and high-carbon steel, and how making alloys produces metals that are more useful than the pure elements ☐

11 I can describe that crude oil is a mixture of a large number of compounds that can be separated into fractions by fractional distillation, using the differences in boiling points between molecules ☐

12 I can state the trends in boiling points, viscosity and flammability, as the mean molecule size in each fraction gets larger ☐

13 I can recognise that a molecule is an alkane (C_nH_{2n+2}) from its structural formula; and draw the structural formulae of methane (CH_4), ethane (C_2H_6), propane (C_3H_8) and butane (C_4H_{10}) ☐

14 I can describe the process of combustion, and describe the polluting effects of carbon, nitrogen and sulfur compounds ☐

15 I can evaluate the economic, ethical and environmental issues surrounding the use of both crude-oil-based fuels and biofuels ☐

16 I can describe how hydrocarbons can be broken down (cracked) to produce alkanes and unsaturated hydrocarbons called alkenes (C_nH_{2n}) that contain a double carbon–carbon bond ☐

17 I can describe how alkenes can be used to make polymers such as poly(ethene) and poly(propene) ☐

18 I can describe some useful applications of polymers; and new uses that are being developed, including biodegradable plastic bags ☐

19 I can describe two ways of making ethanol ☐

20 I can explain how plant material can be processed to produce plant oils ☐

21 I can describe how mixtures of oil and water can be emulsified, and explain how an emulsifier works ☐

22 I can describe how to test if an alkene or a plant oil contains carbon=carbon bonds using bromine water ☐

23 I can describe the structure of the Earth, and explain how tectonic plates cause earthquakes and volcanic eruptions ☐

24 I can describe the chemical content of the Earth's first atmosphere, how it formed, and how it has changed to the atmosphere today ☐

25 I can describe one theory of how the building blocks of life could have been made, and explain why we cannot be sure that this theory is correct ☐

26 I can explain how carbon is stored in rocks, fossil fuels and the sea, and the consequences of increasing the concentration of carbon dioxide in the air ☐

1   I can describe how atoms lose or gain electrons to gain noble gas structures ☐

2   I can describe how atoms join using ionic and covalent bonds to make compounds ☐

3   I can describe the following structures: sodium chloride crystal lattice; metallic lattices; and macromolecules, such as diamond, graphite and silicon dioxide ☐

4   I can explain the different properties of compounds, including electrical conductivity, in terms of their structure ☐

5   I can describe the differences between thermosetting and thermosoftening polymers; how shape memory alloys work; and the benefits and uses of nano science ☐

6   I can describe the atomic structure of the first 20 elements using the periodic table, and use the group number to find the number of outer electrons of the other elements ☐

7   I can calculate the relative atomic mass from data on isotopes, the relative formula mass of a compound, and know that each of these masses in grams is known as a mole of substance and has identical numbers of particles

8   I can calculate the percentage composition of an element in a compound; and calculate the empirical formula of a compound using the masses or percentage of each element present in a sample ☐

9   I can calculate the mass of reactants or products when given data about a reaction from the balanced equation, and can calculate theoretical yields and actual yields ☐

10   I can explain why the theoretical yield in a reaction is not always achieved, and explain that many chemical reactions are reversible ☐

11   I can explain why using instrumental methods, such as gas chromatography and mass spectrometry, provide a quick and reliable analysis of substances ☐

12   I can calculate the rate of a chemical reaction, and know that concentration (and pressure in gases), temperature, particle size, and catalysts all affect the rate of a reaction ☐

13   I can use collision theory to explain how and why the rate of a chemical reaction will change when one variable is altered ☐

14   I can plot, use and interpret rate-of-reaction graphs ☐

15   I can identify exothermic and endothermic reactions and describe their use in hand warmers, sports injury packs and self-heating food applications ☐

16   I can choose the correct acid for making a salt, and describe how to make soluble and insoluble salts ☐

17   I can describe the difference between an alkali and a base ☐

18   I can describe the process of electrolysis of both solutions and molten liquids in terms of electron transfers; and the uses of the products, including electroplating and obtaining sodium and chlorine from molten sodium chloride, and sodium hydroxide, chlorine and hydrogen from a solution of sodium chloride ☐

1 I can describe how the periodic table was developed, and the reasons Mendeleev's version was successful ☐

2 I can explain the trends in reactivity of Group 1 and Group 7 elements in terms of the electronic structures of each element ☐

3 I can describe the properties of transition metals and their compounds in terms of physical properties and chemical properties ☐

4 I can describe how both permanent and temporary hard waters are different from soft waters ☐

5 I can measure the hardness of water using soap solution, and know how to remove both permanent and temporary hardness from water ☐

6 I can describe how water can be purified for drinking, including the use of chlorine to kill microbes. I understand the use of domestic water filters to remove dissolved substances and to kill microbes found in tap water ☐

7 I can explain why fluorine may be added to water for health reasons, and evaluate the economics of distillation as a method of water purification ☐

8 I can calculate the energy change from a chemical reaction, describe if a reaction is exothermic or endothermic, and explain the energy changes in terms of reactant bonds breaking and product bonds forming ☐

9 I can interpret an energy level diagram, and understand that the activation energy is the energy required to break the bonds of the reactants. I can describe the effect of catalysts on activation energy and draw the effect on an energy level diagram ☐

10 I can identify both metal and non-metal ions in solution: using flame tests and sodium hydroxide to identify metal ions; and silver nitrate, hydrochloric acid, and barium chloride solutions to identify non-metal ions ☐

11 I can calculate from titration data the concentration of a solution, and the mass of a product formed ☐

12 I can describe a dynamic equilibrium and explain how changing one condition will affect the proportions of reactants to products ☐

13 I can identify alcohols, carboxylic acids and esters from their formulae or structures ☐

14 I can describe the reactions of alcohols, carboxylic acids and esters ☐

15 I can describe the effect of dissolving alcohols and carboxylic acids in water, and how carboxylic acids will ionise to form weak acids ☐

Group 1	Group 2													Group 3	Group 4	Group 5	Group 6	Group 7	Group 0
						1 H 1 hydrogen													4 He 2 helium
7 Li 3 lithium	9 Be 4 beryllium													11 B 5 boron	12 C 6 carbon	14 N 7 nitrogen	16 O 8 oxygen	19 F 9 fluorine	20 Ne 10 neon
23 Na 11 sodium	24 Mg 12 magnesium													27 Al 13 aluminium	28 Si 14 silicon	31 P 15 phosphorus	32 S 16 sulfur	35 Cl 17 chlorine	40 Ar 18 argon
39 K 19 potassium	40 Ca 20 calcium	45 Sc 21 scandium	48 Ti 22 titanium	51 V 23 vanadium	52 Cr 24 chromium	55 Mn 25 manganese	56 Fe 26 iron	59 Co 27 cobalt	59 Ni 28 nickel	64 Cu 29 copper	65 Zn 30 zinc			70 Ga 31 gallium	73 Ge 32 germanium	75 As 33 arsenic	79 Se 34 selenium	80 Br 35 bromine	84 Kr 36 krypton
85 Rb 37 rubidium	88 Sr 38 strontium	89 Y 39 yttrium	91 Zr 40 zirconium	93 Nb 41 niobium	96 Mo 42 molybdenum	99 Tc 43 technetium	101 Ru 44 ruthenium	103 Rh 45 rhodium	106 Pd 46 palladium	108 Ag 47 silver	112 Cd 48 cadmium			115 In 49 indium	119 Sn 50 tin	122 Sb 51 antimony	128 Te 52 tellurium	127 I 53 iodine	131 Xe 54 xenon
133 Cs 55 caesium	137 Ba 56 barium	139 La 57 lanthanum	178 Hf 72 hafnium	181 Ta 73 tantalum	184 W 74 tungsten	186 Re 75 rhenium	190 Os 76 osmium	192 Ir 77 iridium	195 Pt 78 platinum	197 Au 79 gold	201 Hg 80 mercury			204 Tl 81 thallium	207 Pb 82 lead	209 Bi 83 bismuth	210 Po 84 polonium	210 At 85 astatine	222 Rn 86 radon
223 Fr 87 francium	226 Ra 88 radium	227 Ac 89 actinium																	

C1 Answers
Pages 68–69
Atoms, elements, and compounds

1a copper [1]

b copper carbonate [1]

c It is made of more than one element. [1]

d Malachite contains both rock and copper carbonate [1], so it is a mixture [1].

Inside the atom

1a An atom of phosphorus contains 15 protons [1], 15 electrons [1] and has a mass of 31. [1] It has 16 neutrons. [1]

b i N, As, Sb, or Bi [1]

   ii Cl [1]

   iii Si [1]

c 3 concentric circles with two crosses on inner most circle, eight crosses on middle circle [1] and five crosses on outer circle. [1]

2a

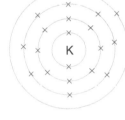

b

[2]

c They both have one outer electron [1], this means that their reactions will be similar [1].

Element patterns

1a by proton number [1]

b periods [1]

c groups [1]

d They have similar properties/characteristics, [1] just like people in the same family. [1]

2a helium (2), neon (2,8), argon (2,8,8), krypton (2,8,18,8), xenon (2,8,18,18,8), or radon (2,8,18,32,18,8), [1] for choosing a noble gas and with electronic structure [1]

b They have a filled outer electron shell [1], that makes them unreactive [1].

Combining atoms

1a a shared pair of electrons [1]

b

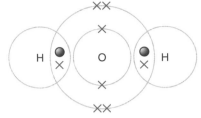

One mark for showing a pair of electrons between O and H atom, [1] for eight electrons on the outer shell of O atom.

c 2,8,1 [1]

d It loses its outer electron [1], and loses a negative charge, so it becomes a positively charged ion.[1]

e The chlorine gains the electron lost by the sodium[1], and becomes a negative ion [1], which is then attracted to the sodium ion.[1]

2a a double covalent bond. [1]

b One mark for showing two pairs of electrons joining each atom together; one mark for showing eight electrons on each outer shell.

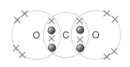

Pages 70–71
Chemical equations

1a Carbon dioxide. [1]

b i 3 [1]

   ii 2 [1]

2a $CuCO_3 + 2HCl \rightarrow CuCl_2 + CO_2 + H_2O$ [1]

b $CuCO_3 + 2NaOH \rightarrow Cu(OH)_2 + Na_2CO_3$ [1]

Building with limestone

1 Social advantages: more employment, more money to pay for community facilities. [1]

Social disadvantages: pollution effects on local population (not on wildlife). [1]

Economic advantages: more money in the local economy, better roads and railways. [1]

Economic disadvantages: reduction in tourism. [1]

To gain full marks you must mention both advantages and disadvantages.

2a Carbon dioxide [1] dissolves in the rain to make carbonic acid [1] that dissolves the limestone [1].

b The water with the dissolved limestone evaporates [1] and the carbon dioxide from the air makes calcium carbonate [1] that then becomes stalactites and stalagmites.

Page 84

The atmosphere and life

1a There are no fossil records of early life/It was too small to leave traces [1].

b Seawater allows all the compounds to be able to meet each other/Ultraviolet light could not damage or kill the emerging life. [1]

c All living things/proteins are made from amino acids [1]. If we know how to make amino acids then it is a possible route to making life [1].

2a ammonia (NH_3), methane (CH_4) and hydrogen (H_2) (all three for the mark) [1]

b sugars/ribose sugar [1] and amino acids [1]

c It didn't make a living thing [1], it only made building blocks/compounds that are present in living things [1].

Carbon dioxide levels

1a i A = photosynthesis [1]

ii B = burning [1]

iii C = death and decay [1]

iv D = respiration [1]

b Both plants and animals carry out respiration. [1]

c The carbon dioxide produced by burning, and given off to the air [1], has recently been removed from the air by photosynthesis [1].

2a It's a fuel made from compounds containing a lot of carbon atoms. [1]

b Fossil-fuel carbon has been locked up for millions of years and released quickly [1]; whereas biofuel carbon has only recently been removed from the air and is returned as quickly as photosynthesis takes place [1].

c Carbon dioxide is an acidic gas, so it will affect the pH of the water [1]. This change in pH may dissolve sea shells/make it hard for plants to photosynthesise/kill some living organisms [1].

Page 85

Extended response question

5 or 6 marks:

There is a clear, balanced and detailed description of the positive and negative environmental impacts involved with increasing the percentage of biofuels, with 5–6 points from the examples given. The answer shows almost faultless spelling, punctuation and grammar. It is coherent and in an organised, logical sequence. It contains a range of appropriate or relevant specialist terms used accurately.

3 or 4 marks:

There is some description of the positive and negative environmental impacts involved with increasing the percentage of biofuels, with 3–4 points from the **examples** given. There are some errors in spelling, punctuation and grammar. The answer has some structure and organisation. The use of specialist terms has been attempted, but not always accurately.

1 or 2 marks:

There is a brief description of the environmental impacts involved with increasing the percentage of biofuels, with 1–2 points from the **examples** given. The spelling, punctuation and grammar are very weak. The answer is poorly organised with almost no specialist terms and/or their use demonstrates a general lack of understanding of their meaning.

Possible points to make:
positive
- carbon neutral
- growing plants will remove carbon dioxide from atmosphere
- less crude oil will be needed
- less risk of oil spills if less oil needed
- biofuels are natural, and spillages will be easier to deal with.

negative
- will not reduce carbon emissions as still burning carbon dioxide
- more land will be needed to grow crops
- rainforest may be destroyed
- animal habitats may be disrupted or destroyed.

C2 Answers

Pages 86–87

Investigating atoms

1a

particle	Relative mass	Relative charge
proton	(i) 1	+1
(ii) neutron	1	(iii) none
electron	(iv) very small	-1

(for 1 mark each)

b i 14 [1]

ii 14 [1]

iii 2,8,4 [1]

2a alpha particles [1]

b thin layer of gold foil [1]

c to show where the alpha particles are deflected [1]

d D: undeflected, have missed the nucleus showing the atom is mainly empty [1]. E: those deflected forward have had their direction of travel influenced by a central nucleus/passing close to the nucleus [1]. C: those deflected backwards have been bounced back from a central nucleus [1].

Mass number and isotopes

1a An isotope is an atom of an element that has a different mass to other atoms. [1]

b Potassium-40 has one [1] more neutron [1].

c $\dfrac{(90 \times 39) + (10 \times 40)}{100} = \dfrac{3910}{100} = 39.1$ [2]

Another answer, but showing correct method, [1].

2 Measure the carbon-14 present in the sample [1], compare it with how much should be present today/ use half-life of carbon-14 to find out how old the specimen is [1].

Compounds and mixtures

1a an aqueous or water solution [1]

b 137 + (35.5 × 2) [1 mark] = 208 [1 mark]

c 208 grams [1]

d (27 × 2) + ((32+(16 × 4)) × 3) [1 mark] = 342 [1 mark]

Electronic structure

1a 2,8,8,2 [1] for diagram
[1] for written structure

2,8,8,2

b 2 [1] for diagram
[1] for written structure

2

c 2,8,3 [1] for diagram
[1] for written structure

2,8,3

d 2,7 [1] for diagram
[1] for written structure

2,7

2a i 7 [1]

ii 2 [1]

b It tells you its reactivity/chemistry/how it will react. [1]

c They have full outer electron shells [1] so there are no electrons to share / gain or lose / to react. [1]

Pages 88–89

Ionic bonding

1a high melting or boiling point [1] and conducts in solution but not when solid [1]

b i Either a diagram showing 2,8,8 configuration or a statement that it is 2,8,8. [1]

ii Either a diagram showing 2,8 configuration or a statement that it is 2,8. [1]

c The potassium atom loses its outer electron to the fluorine atom, making two ions [1]. The oppositely charged ions then attract each other [1] [with an electrostatic attraction].

2a When solid, the sodium and chloride ions are not free to move [1]; when dissolved, they are free to move and so can conduct the electric current [1].

b When molten, the sodium and chloride ions are free to move. [1]

Alkali metals

1a sodium + water → sodium hydroxide + hydrogen [1]

b They all have a single outer electron [1], which they lose to become 1+ ions. [1]

c It would react very violently [1] as it is lower in the group and so more reactive than potassium [1].

2a They are less reactive so easier to extract. [1]

b by using electrolysis [1]

c Electricity was only discovered then. [1]

Halogens

1a

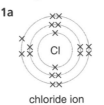

chloride ion

[1]

b 1– [1]

c argon [1]

2a The chlorine in sodium chloride is reacted with the sodium [1]; this makes a new substance with different properties [1].

b Polyvinyl chloride has different properties from sodium chloride and doesn't dissolve in water. [1]

c Enough is used to kill bacteria, but not enough to kill people / people are larger than bacteria and need far more chlorine to be killed. [1]

ii The rate will be faster as the particles are moving faster so will produce more reacting collisions [1]; height will be the same as there are no more particles to react than before [1].

2a Heat energy provides activation energy for the first atoms to burn [1], subsequent heat energy produced from burning magnesium provides activation energy for more atoms to burn [1].

b Increasing temperature provides more energy to the particles [1], this means that more particles will possess enough energy to react and this is the activation energy [1].

Pages 98–99

Concentration

1a A [1]

b C [1]. The curve is lowest/takes longest time to reach the end point/plateau [1].

c This is the maximum amount of hydrogen gas that can be produced from the reactants [1]. Changing the rate only changes the time it takes to make the products, not the quantity of products [1].

d 3.6 min [1]

2a draw tangent at 2, and take readings, for example:
$$\frac{7.4 - 4.8}{3 - 1} = 2.6/2 = 1.3$$

accept answer between 1.1 and 1.5. (1 mark for tangent figures, and 1 mark for answer between 1.1 and 1.5)

b 1.8 minutes [1]. Doubling concentration will halve the time for the reaction to complete. [1]

Size matters

1a iron filings [1]

b The iron filings have the smallest particles [1], so will react fastest as there are more particles available to react at the same time [1].

c The nail did not react [1]. There were no brown bits on the mat at the end/it was still the same size at the end [1].

2a Sprayed fuel forms drops with a larger surface area [1], so it is easier for the oil molecules to burn [1].

b The smaller drops of oil in the spray burn quicker [1] this releases heat energy faster, raising the temperature quickly. [1]

Clever catalysis

1a Reaction is five time faster [1 mark] as 250/50 = 5 [1 mark]

b The reaction rate was quicker with manganese(IV) oxide present [1] and there was no change in the mass of manganese(IV) oxide during the experiment [1].

c It could have lowered the activation energy; or it could have formed an intermediate that reacted easily, [1] so making it easier to release the oxygen [1].

2a Metals like lead can coat the catalyst's surface [1];

this would stop the catalyst working [1].

b Although there is no lead present there may be other substances that could coat the catalyst's surface [1], these may be present in air or the fuel [1].

Controlling important reactions

1a nitrogen + hydrogen ⇌ ammonia [1]

b It is very slow. [1]

c Increasing the temperature increase the rate of a reaction [1], so to make the ammonia quickly 450 °C is used [1]

d Increasing the pressure forces the molecules closer together [1], this means there will be more collisions so a quicker reaction [1].

2a The ammonia produced can be easily separated from the unreacted gases [1], which can then be recycled so eventually all the reactants will become products [1].

b Any three from: high temperature and pressure make the reaction quick [1], but high temperature gives a poor yield but quick reaction time [1]; high pressure increases yield, but is very costly[1]; the combination chosen gives best yield for least time and cheapest energy cost [1].

Pages 100–101

The ins and outs of energy

1a endothermic [1]

b The products need more energy than the reactants [1], so energy is taken from the solution so the temperature falls [1].

c quicklime and water [1]

d The quicklime and water have more stored energy than the products they make [1], the surplus energy is released as heat energy [1].

2a a reaction that can go in either direction [1]

b the reaction from left to right in the written equation [1]

c The backward reaction is the exact opposite of the forward reaction [1], whatever energy change happens with the forward reaction, the exact opposite will happen with the backward reaction [1].

Acid–base chemistry

1a An alkali can dissolve in water, a base cannot. [1]

b sodium nitrate and water [1]

c copper chloride [1] and carbon dioxide and water [1]

2a The universal indicator turned blue showing it is an alkaline solution. [1]

b NH_4^+(aq) [1] and OH^- (aq) [1]

c The hydroxide ions react with the hydrogen ions from the acid to form water. [1]

Making soluble salts

1a sulfuric acid [1]

b **i** warm the mixture [1]

ii When no more cobalt oxide dissolves [1], check the pH of the solution [1].

c Filter the solution. [1]

d Evaporate the solution [1] until crystals appear[1].

2a fizzing/a gas being produced [1]

b The calcium sulfate produced did not dissolve in the solution [1], it coated the calcium carbonate preventing the sulfuric acid from reaching the calcium carbonate so the reaction stopped. [1]

Insoluble salts

1a calcium carbonate and sodium chloride [1]

b Filter the solution. [1]

c Rinse the contents of the filter paper with water. [1]

2a i A yellow precipitates is formed. [1]

ii The precipitate is silver iodide. [1]

b white [1]

c $2AgNO_3(aq) + MgCl_2(aq) \rightarrow Mg(NO_3)_2(aq) + 2AgCl$ (1 mark for the unbalanced chemical equation, 1 mark for balancing the equation)

Page 102

Ionic liquids

1a electrolysis [1]

b Dichromate ions are negatively charged [1] and will be attracted by the opposite charge of the positive electrode [1].

c The ion will gain an electron [1] becoming an atom of potassium or react with the water becoming potassium hydroxide [1].

2a Electricity is a flow of electrons, the potassium ion gains an electron at the negative electrode [1], the dichromate ions lose (two) electrons at the positive electrode; this allows for a flow of electrons in the circuit [1].

b In AC, the current rapidly changes polarity [1] which means that the ions will be attracted first one way, then the other, with the result that they will not move from the original position [1].

Electrolysis

1a the positive electrode [1]

b It loses two electrons [1] and dissolves in the solution [1].

c to allow copper to be deposited on the negative electrode [1]

d the impurities from the copper [1]

2a It only shows the reaction at one electrode, which is only half the reaction. [1]

b $2Cl^-(\ell) \rightarrow Cl_2(g) + 2e^-$ [1 mark]

c Oxidation is the loss of electrons: and the chloride ions lose electrons. [1]

d $2NaCl(\ell) \rightarrow 2Na(\ell) + Cl_2(g)$ [1 mark]

Page 103

Extended response question

5 or 6 marks:
There is a clear, balanced and detailed description of the positive and negative reasons for the use of brine rather than molten sodium chloride with 5–6 points from the **examples** given. The answer shows almost faultless spelling, punctuation and grammar. It is coherent and in an organised, logical sequence. It contains a range of appropriate or relevant specialist terms used accurately.

3 or 4 marks:
There is some description of the positive and negative reasons for the use of brine rather than molten sodium chloride with 3–4 points from the **examples** given. There are some errors in spelling, punctuation and grammar. The answer has some structure and organisation. The use of specialist terms has been attempted, but not always accurately.

1 or 2 marks:
There is a brief description of the reasons for the use of brine rather than molten sodium chloride with 1–2 points from the **examples** given. The spelling, punctuation and grammar are very weak. The answer is poorly organised with almost no specialist terms and/or their use demonstrating a general lack of understanding of their meaning.

possible points include:
positive

- it can be carried out at room temperature so saves energy
- hydrogen is produced at a low temperature
- so it will be harder to ignite
- sodium hydroxide is produced directly, with molten reagents a further reaction is needed.
- with brine the sodium immediately reacts with the water solution to make NaOH.

negative

- with molten sodium chloride the sodium made has to react with water
  $2Na + 2H_2O \rightarrow 2NaOH + O_2$
- the more processes used the costlier it will be.

C3 Answers

Pages 104–105

Ordering elements

1a Any two from: Group 7 is at the left-hand side; Cu is included in the Alkali metals and it shouldn't be; no Group 0 elements. [2]

b Any two from: the lower down the table the less the arrangement reflected the Chemistry; cobalt was not like chlorine; there was no definite pattern; a box had two elements in it. [2]

c He left spaces for undiscovered elements [1] and did not keep to increasing mass order when the chemistry suggested otherwise [1].

2a germanium [1]

b The prediction was 72, the mass is 73. [1]

c It showed his predictions were reliable. [1]

d Tellurium should be under bromine, but iodine shows the same chemistry as bromine so Mendeleev swapped them. [1]

e Group 0 / the noble gases [1], they had not been discovered [1].

The modern periodic table

1a Group 1 [1] because it has a single outer electron [1].

b

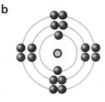

[1]

c It is easy for sodium to lose its outer electron, [1] argon has a full outer electron shell and is unreactive as a result [1].

d Both need to gain a single electron, so their chemistry will be the similar. [1]

Group 1

1a potassium [1]

b Potassium's outer electron is easiest to lose [1] because it is furthest from the positive nucleus [1] and there are more electron shells to reduce the effect of the attraction [1].

c Any three from: it would float; move around and fizz/see bubbles; turn into a round ball/turn the water purple/blue. [3]

d i sodium hydroxide [1]
ii 23 + 1 + 16 = 40 [1]

2 They are far too reactive [1] and would probably explode [1].

Transition metals

1a Their compounds are coloured. [1]

b Iron can form two ions, [1] Fe^{2+} and Fe^{3+} [1].

c i a substance that speeds up the rate of a reaction, and is unchanged by the reaction. [1]

ii to reduce air pollution [1], to convert nitrogen oxides back to nitrogen to reduce acid rain [1]

2a The greater the cost, the lower the abundance/The more abundant, the cheaper the cost. [1]

b osmium and rhodium [1]

c It may be harder to find [1] it may be more costly to extract/reduce to metal [1].

d Iron is so much cheaper [1] that the cost savings are more important than the reduced catalytic effect [1].

e Economically, the platinum and rhodium will be worth something [1]; environmentally, it reduces the need to mine or extract the metal, both of which would be polluting [1].

Pages 106–107

Group 7

1a 7 [1]

b 1- [1]

c Fluorine needs one electron to fill the outer shell [1]. The shell is very close to the nucleus, so there is a great attraction for electrons [1]. There are very few shells to shield the nucleus's positive charge [1].

2a

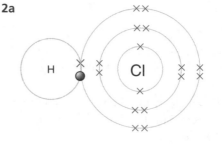

b The molecule splits into ions [1], one of the ions is H^+ (aq) which is the ion that causes acidity [1].

c In the gas, the molecules are far apart and uncharged [1]; as ions in solution, the particles are charged and can carry the current/accept electrons and lose electrons [1].

Hard and soft water

1a A and E [1]

b C [1] it needs less soap after boiling, and produces no scum after boiling [1]

c D [1] it needs less soap after boiling, but still produces scum after boiling, so it must contain both [1]

d Any two from: add sodium carbonate [1], distil the water [1], use an ion exchange column [1].

2a (ℓ) [1 mark]

b calcium hydrogencarbonate [1]

c It removes the calcium ions from the solution [1] that react with the soap molecule to form scum [1].

Safe drinking water

1a It removes particles/solids/chlorine from the water. [1]

b Ion resins exchange metal ions by absorbing them [1] and releasing H^+ ions instead.

c The silver kills any microbes in the water. [1]

2a removal of salt from sea water [1]

b The seawater is heated to boiling point [1], the water then turns to gas [1] leaving the salt behind [1], the pure water then condenses in the condenser [1].

c The cost of the energy needed is too large to make it economic. [1]

Energy from reactions

1a energy = m x c x ΔT [1 mark] so

200 x 4.2 x 7 = 5 880J [1 mark]

b The thermometer is touching the beaker and measuring the temperature of the beaker [1]. Suspend the thermometer in the water [1] to correct the problem.

c There were lots of heat losses to the environment [1], the specific heat capacity took no account of heating the apparatus [1].

2a (12 x 2) + (1 x 6) + (16 x 1) = 46 [1]

b 46/2 g [1] x 5 880 = 135 240 J/mol [2]

Pages 108–109

Energy from bonds

1a releases heat energy [1]

b More energy is released when the product bonds are formed [1] than was needed to break the reactant bonds [1].

c C_2H_6 has 6 C–H bonds, so 6 × 413 = 2478, it has 1 C–C bond at 346, so C_2H_6 needs 2824 kJ. As 2 moles are used in the equation, 2824 × 2 = 5648 kJ. 7 O=O bonds = 7 × 496 = 3472 + 5648 = 9120 kJ [1]. Products have 8 × C=O bonds = 8 × 743 = 5944, 12 O–H bonds, 12 × 463 = 5556, so products = 11500 [1] So 9120 – 11500 = –2380, but 2 moles of C_2H_6 was burnt, so –2380/2 = –1190 kJ/mol [1]

2a 1662/4 = 415.5 kJ mol⁻¹

b In methane there are four C–H bonds. Each bond needs different energies to break. [1] The value of 415.5 kJ/mol is the average for the four different values. [1]

Energy saving chemistry

1a $2H_2(g) + O_2(g) \rightarrow 2H_2O(\ell)$ [1 mark]

b by the electrolysis [1] of water [1]

c The electricity needs to be produced in power stations [1] that burn fossil fuels [1], this causes the acid rain.

d Hydrogen is explosive, so crashed cars may explode if the fuel tank is burst. [1]

2a It removes electrons from the hydrogen. [1]

b It adds electrons to hydrogen ions that react with the oxygen. [1]

c They are expensive [1] and also unreliable [1].

Analysis – metal ions

1a i copper [1]
　　 ii nickel or iron (II) [1]
　　 iii iron(III) [1]

b Group 1 [1]

c aluminium [1]

Analysis – non-metal ions

1a i B [1]
　　 ii E [1]
　　 iii D [1]

b Add barium chloride solution [1]. It should produce a white precipitate [1].

2a Barium sulfate is insoluble [1], this means it cannot be absorbed into the body even if drunk [1].

b Barium sulfate absorbs X-rays [1], allowing X-ray photographs to clearly show your digestive system [1].

Pages 110–111

Measuring in moles

1a 59 + (35 × 2) = 129 [1], 129/129 = 1 mole [1]

b 56 + (35 × 2) = 126 [1], 12.6/126 = 0.1 mole [1]

c 12 + (1 × 4) = 16 [1 mark], 48/16 = 3 moles [1]

2a 64.5/129 = 0.5 moles [1 mark] 0.5 × 1000/500 = 1 mol/dm³ [1]

b 25.2/126 = 0.2 moles [1], 0.2 × 1000/200 = 1 mol/dm³ [1]

c HCl = 1 + 35 = 36, 7.2/36 = 0.2 moles [1], 0.2 × 1000/500 = 0.4 mol/dm³ [1]

Analysis – acids and alkalis

1a 20.2 [1] (remember to omit the rough trial, and only record the value to the same number of decimal places as the measurements were)

b 0.1 × (20.2/1000) = 0.00202 moles (1 mark for equation, 1 mark for correct answer)

c 0.00202 moles [1]

d 0.00202 × 1000/25 = 0.0808 mol/dm³ [1 mark for equation, 1 mark for correct answer]

2a 1 × (24.5/1000) = 0.0245 moles [1 mark for equation, 1 mark for correct answer]

b From the equation, the reacting ratios are 1 : 2 [1 mark], so 0.0245 × 2 moles of ammonia are present = 0.0490 [1 mark]

c $(NH_4)_2SO_4$ = (14 × 2) + (1 × 8) + 32 + (16 × 4) = 132 [1 mark], 132 × 0.0245 = 3.23 g [1]

Dynamic equilibrium

1a The oxygen mass will increase [1]. The reaction is exothermic, so increasing the temperature will move the equilibrium to the left (the backward reaction) [1].

b Nothing [1] a catalyst only increases the rate, it doesn't affect the proportions made [1].

2 The reactants mass will decrease [1], increasing the pressure makes the equilibrium move to reduce the number of particles/molecules, so the forward reaction is favoured making more products [1].

Making ammonia

1a 200 °C [1]

b it increases [1]

c 60% [1]

2a 26–28% (1 mark for a figure between these numbers)

b Any four points from:

- increasing temperature, increases yield.
- increasing pressure, increases yield.
- increasing pressure is costly.
- increasing temperature slows the reaction.
- conversion percentage does not matter as the unreacted nitrogen and hydrogen are recycled and eventually become ammonia.
- it is a compromise between speed of reaction, cost and yield. (4 marks)

Page 112

Alcohols

1a alcohol [1]

b water [1] carbon dioxide [1]

2a 1 mark for removing two H atoms and replacing with an oxygen, 1 mark for a double bond to the carbon, and all C atoms only having 4 bonds.

b ethanoic acid [1]

Carboxylic acids

1a It produces steam that can condense/water [1].

b an ester [1]

c it can act as a flavouring[1] and scent[1]

2a sodium ethanoate [1], carbon dioxide [1], water[1]

b weak acids [1]

c All the hydrochloric acid molecules dissociate in water [1] but only some of the ethanoic acid molecules dissociate [1].

Page 113

Extended response question

5 or 6 marks:
There is a clear, balanced and detailed description of the experiment and an appropriate risk assessment, with 5–6 points from the examples given.

The answer shows almost faultless spelling, punctuation and grammar. It is coherent and in an organised, logical sequence. It contains a range of appropriate or relevant specialist terms used accurately.

3 or 4 marks:
There is some description of the experiment that may include a risk assessment, with 3–4 points from the examples given.

There are some errors in spelling, punctuation and grammar. The answer has some structure and organisation. The use of specialist terms has been attempted, but they have not always been used accurately.

1 or 2 marks:
There is a brief description of the experiment that may include a risk assessment, with 1–2 points from the examples given.

The spelling, punctuation and grammar are very weak. The answer is poorly organised with almost no specialist terms and/or their use demonstrates a general lack of understanding of their meaning.

Points to make:

- describes a known volume or mass of water to heat
- weighs burner before and after experiment
- limits temperature rise to no more than 10 °C, or time to burn, e.g. 5 minutes
- suggests some degree of shielding/insulation of the apparatus
- states that

  Q = mass in kg × specific heat capacity (4.2) × temperature rise

- identifies need to scale up the mass burnt to 46 g by either: mass used in g/46 × temperature rise

OR

mass used in g/46 × energy released

- has a risk assessment with hazard, risk, and control measure.